THE HUMAN GEOGRAPHY OF EASTERN EUROPE

This is a major review of the human geography of Eastern Europe since the Communist takeover in the immediate post-war period. The volume assesses the major points of difference and similarity between individual countries, analysing the extent to which central planning and Soviet influence have produced uniformity and the extent to which regional identities have been maintained.

Dividing his work into two major sections, David Turnock first explores trade, transport and energy, covering the foundation role of Comecon, the restructuring of transport services, and the establishment of the nuclear programme. The second section offers important information on housing policies, settlement planning, and environmental protection.

Throughout, the volume discusses how the human geography of Eastern Europe has been redrafted in the Communist period, seeing points of continuity as well as radical new departures. Illustrated with original cartography, this wide-ranging text will be a standard reference for all students of Eastern European geography.

David Turnock is Reader in Geography at the University of Leicester.

THE HUMAN GEOGRAPHY OF EASTERN EUROPE

David Turnock

ROUTLEDGE
London and New York

First published 1989 by Routledge
11 New Fetter Lane, London EC4P 4EE
29 West 35th Street, New York, NY 10001

© 1989 David Turnock

Printed and bound in Great Britain by
Biddles Ltd, Guildford and King's Lynn

British Library Cataloguing in Publication Data

Turnock, David.
    The human geography of Eastern Europe.
    1. Europe. Eastern Europe. Human geography
    features.
    I. Title
    304.2'0947

    ISBN 0-415-00469-1

Library of Congress Cataloging in Publication Data

Turnock, David.
    The human geography of Eastern Europe.
    Bibliography: p.
    Includes index.
    1. Anthropo—geography—Europe, Eastern.
2. Europe, Eastern—Economic conditions—1945–   64445
I. Title.
GF645.E92T87   1989   304.2'0947   88-31679
ISBN 0-415-00469-1

CONTENTS

v

LIST OF FIGURES

LIST OF TABLES

ACKNOWLEDGEMENTS

Eastern Europe is already well served by geography texts in the English language. Although the number of books is not great in comparison with other regions of the world there is useful complementarity and the reader can choose between books which provide detailed information on individual countries with only limited coverage of general themes relevant to the region as a whole (Osborne 1967; Pounds 1969) and works which concentrate on the broader issues to the partial or complete exclusion of individual national profiles (Mellor 1975; Turnock 1978). However all the books available are now several years out of date and cannot offer much guidance on the important preoccupations of the present and the immediate future. Equally the literature available in the journals for the 1980s cannot be summarized and referenced for the reader's convenience. It is the purpose of the present work to provide an update concentrating on themes relevant to the late 1980s with general outlines supplemented by examples and case studies wherever possible. The work complements the author's previous books on The Historical Geography of Eastern Europe published by Routledge in 1988.

It was originally intended that the whole work would be published in one volume but it has proved necessary to divide the material into two separate publications. A companion volume entitled The Political and Economic Geography of Eastern Europe examines the state system and the principal economic sectors while this work covers what might be regarded as the infrastructure: trade, transport, energy, settlement patterns, and conservation. Both books are divided into two sections, each comprising four chapters. The first chapter in each section provides a general review and a contextual essay for the more specialized studies which follow. Many examples (and several maps) are drawn from the author's own research on Romania. While the emphasis on one country may appear excessive it does allow the exemplification of many aspects of central planning without widespread copying of the work of other writers.

Because there are other textbooks available to provide basic groundwork and to give detailed coverage of the events of the early post-war period the referencing of articles is generally restricted to work which has appeared since 1970 (although in the case of books the more valuable items of the 1945-70 period are mentioned). Referencing is also limited to works in the English language, since a fully comprehensive bibliography would be excessively lengthy and probably counter-productive for the majority of readers. It is often necessary to refer to particular parts of certain countries and the word 'region' is reserved for the official administrative regions (to avoid the complexities in using the various national labels like 'Bezirke' in the GDR, 'judet' in Romania, or 'wojewodztwo' in Poland).

Thanks are due to all the authors whose research is acknowledged in the pages that follow: without such analyses and commentaries works of synthesis would be virtually impossible. It is also a pleasure to thank Peter Sowden and Alan Jarvis at Routledge for their encouragement over the project and forbearance over the delays in delivering the typescript.

Leicester, July 1988   David Turnock

INTRODUCTION

Eastern Europe has never been prominent in the minds of English speakers, and westerners generally for that matter, but the rise of Russia to superpower status in the context of a Marxist ideology has converted the heart of Europe into a geopolitical watershed and given a new significance to the states lying immediately to the east of this divide (Figure 0.1). A substantial part of what used to be called 'Central Europe' has emerged as 'Eastern Europe' with coherence arising from government by communist parties conforming closely to the Soviet model. The fundamental characteristics of a socialist system of government are difficult to identify beyond appreciation of the Marxist inspiration of radical action by elites seeking to overcome the perceived evils of capitalism. The Soviet model of 'rapid industrialisation, centrally planned within a system of state ownership of the means of production' and limited respect for consumption benefits and other human rights has been widely adopted by communist parties which have voluntarily adopted elements of the system or by others which have experienced Soviet coercion. (1) The visitor will certainly be impressed by the ability of monopoly parties to project themselves through propaganda and architecture; by the direct control exerted over the economy through state-owned industry and through the organization of most of the agriculture and services on a co-operative basis; and by the control of the individual through restrictions on international travel and a preference for investment in communal facilities rather than enhanced allocation of resources to individual families through higher wages and welfare benefits.

1

Figure 0.1 The countries of Eastern Europe today

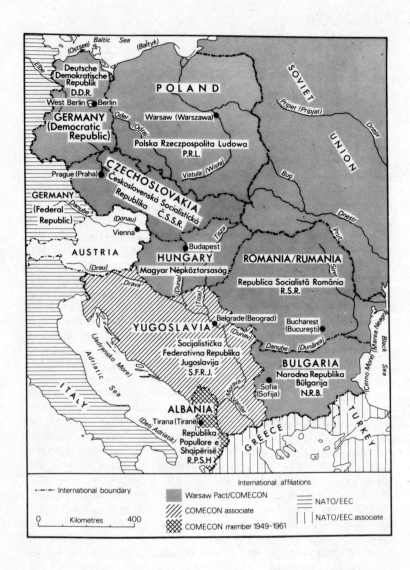

Source: World Atlases.

Writers are often preoccupied with the socialist character of Eastern Europe. It follows in part from an interest in regional geography and the responsibility of portraying an area as significantly, if not uniquely, distinct from others. It may also follow from the reaction to modelling in geography with its underlying assumption of uniformity: socialist Eastern Europe could be portrayed as being fundamentally different from the capitalist west. Differences can however be overstated and it is important to look beyond the simplistic formula that all modern developments are distinctly socialist because they are conditioned by socialist governments. 'Certainly the idea of a socialist landscape is intriguing but its establishment will rest on continued research on specific forms,' their origin and change through time.' (2) Equally, assertions about the inherent superiority of the socialist system should be subject to scrutiny. Revolutionary commitment by ruling groups should not conceal the possibility of more sustained development by governments which do not claim any emotive ideological attachment. Excessive copying of the Soviet model by small nation states which cannot isolate themselves from the wider world market indefinitely has resulted in a cumbersome decision-making system which has difficulty in working towards international specialization with the requirement of constant adaptation to changing world market trends. But such shortcomings may be regarded primarily as a Soviet responsibility since the Russians have imposed constraints on economic reform without being prepared to maintain financial concessions in compensation. Autarky has certainly been broken down since the Khrushchev era when Comecon began to work for greater specialization and closer contacts with the non-communist world have, on the whole, satisfied an important precondition for further progress. Realizing the potential is an issue which lies at the root of Eastern Europe's economic crisis today. With the Gorbachev era in the Soviet Union it seems that greater realism will prevail in the years ahead.

It cannot be overlooked that socialist policies have been applied to a terrain with its own specific characteristics (Figure 0.2). With an area of 1.27 million square kilometres Eastern Europe is only 5.7 per cent the size of the Soviet Union, 13.6 per cent of the USA but is four times larger than the British Isles. In terms of north-south distances the maximum and minimum figures are 1.65 and 1.05 thousand kilometres respectively. Comparable figures for east-west

3

Figure 0.2 Physical background

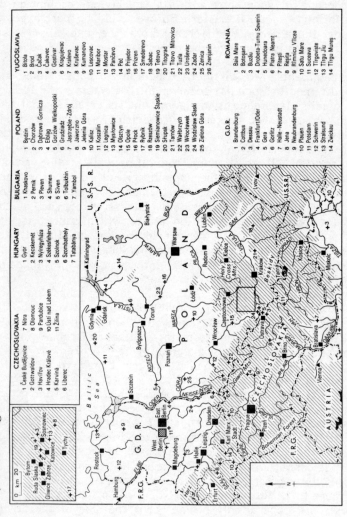

CZECHOSLOVAKIA		HUNGARY	BULGARIA	POLAND		YUGOSLAVIA	
1 České Budějovice	7 Nitra	1 Gyor	1 Khaskovo	1 Bedzin	14 Olsztyn	1 Bitola	14 Peč
2 Gottwaldov	8 Olomouc	2 Kecskemét	2 Pernik	2 Chorzów	15 Opole	2 Brod	15 Prijedor
3 Havířov	9 Pardubice	3 Nyíregyháza	3 Pleven	3 Dąbrowa Gornicza	16 Plock	3 Čačak	16 Prizren
4 Hradec Králové	10 Ústí nad Labem	4 Székesfehérvár	4 Shumen	4 Elbląg	17 Rybnik	4 Čakovec	17 Smederevo
5 Karviná	11 Žilina	5 Szolnok	5 Sliven	5 Gorzów Wielkopolski	18 Rzeszów	5 Gostivar	18 Šabac
6 Liberec		6 Szombathely	6 Tolbukhin	6 Grudziądz	19 Siemianowice Śląskie	6 Kragujevac	19 Tetovo
		7 Tatabánya	7 Yambol	7 Jastrzębie - Zdrój	20 Słupsk	7 Kralevo	20 Titograd
				8 Jaworzno	21 Tarnów	8 Kruševac	21 Titovo Mitrovica
				9 Jelenia Góra	22 Wałbrzych	9 Kumanovo	22 Tuzla
				10 Kalisz	23 Wrocławek	10 Lescovac	23 Urošévac
				11 Koszalin	24 Wodzisław Śląski	11 Maribor	24 Zadar
				12 Legnica	25 Zielona Góra	12 Mostar	25 Zenica
				13 Mysłowice		13 Pančevo	26 Zrenjanin

G.D.R.		ROMANIA	
1 Brandenburg	8 Jena	1 Baia Mare	8 Resiţa
2 Cottbus	9 Neubrandenburg	2 Botoşani	9 Rimnicu Vîlcea
3 Dessau	10 Plauen	3 Buzău	10 Satu Mare
4 Frankfurt/Oder	11 Potsdam	4 Drobeta-Turnu Severin	11 Suceava
5 Gera	12 Schwerin	5 Hunedoara	12 Tirgoviste
6 Görlitz	13 Stralsund	6 Piatra Neamţ	13 Tirgu Jiu
7 Halle - Neustadt	14 Zwickau	7 Piteşti	14 Tirgu Mureş

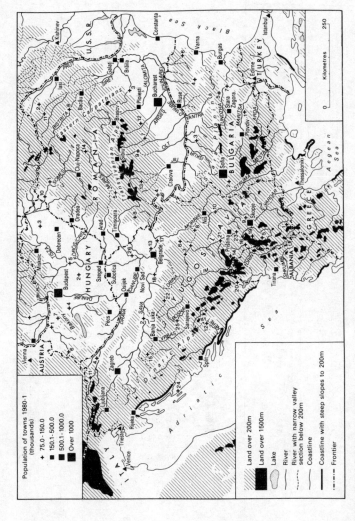

Source: World Atlases.

Table 0.1 Relief and land use'

	Area distributed according to altitudinal zones (thousand square km)						
	Total	Below 200m	200-499m	500-999m	1,000-1,499m	1,500-1,999m	Above 1,999m
Czechoslovakia	127.9	15.1	68.4	40.9	3.0	0.4	*
GDR	108.2	74.6	22.8	10.8			
Hungary	93.0	77.6	14.8	0.6	*		
Poland	312.7	234.2	69.1	9.1	0.3		
North	641.8	401.5	175.1	61.4	3.4	0.4	*
Albania	28.7	7.2	6.2	7.5	5.6	1.9	0.4
Bulgaria	110.9	35.2	38.5	23.2	9.6	3.5	1.2
Romania	237.5	103.8	63.9	45.9	17.8	5.1	1.0
Yugoslavia	255.8	71.8	72.0	70.1	34.0	6.9	1.0
South	632.9	218.0	180.6	146.7	67.0	17.4	3.6
Eastern Europe	1,274.7	619.5	355.7	208.1	70.4	17.8	3.6

Table 0.1 continued

| | Land use (thousand square km) Agriculture | | | |
	Arable	Pasture	Forest	Other
Czechoslovakia	51.7	16.9	45.4	13.9
GDR	50.2	12.3	29.4	16.3
Hungary	53.3	12.9	15.9	10.9
Poland	149.2	40.3	86.9	36.3
North	304.4	82.4	177.6	77.4
Albania	6.0	6.2	13.5	3.0
Bulgaria	43.8	18.1	38.4	10.6
Romania	105.0	44.6	63.4	24.5
Yugoslavia	78.8	63.4	92.6	21.0
South	233.6	132.3	207.9	59.1
Eastern Europe	538.0	214.7	385.5	136.5

Sources: J. Breu (1970) Atlas der Donauländer, Vienna: Österreichisches Ost- und Sudosteuropa Institut, and statistical yearbooks.

Note: * Less than 0.1

distances are 1.25 and 0.65 respectively. Some parts of the region are as much as 600 kilometres from the sea although navigable rivers, to say nothing of modern infrastructure, have reduced the significance of this apparent isolation. Contrasts in relief are spectacular with Alpine-type fold mountain chains (most notably the Carpathian and Dinaric systems) and Hercynian plateau country of Bohemia and adjacent areas contrasting with the North European Plain and other extensive lowlands in Lower and Middle Danubia. In Eastern Europe 23.2 per cent of the land lies above 500 m but the proportion is much higher in the south (37.1 per cent) than in the north (11.6); it is in the south where the alignment of mountain chains achieves particular significance for cultural development through the influence exerted on the thrust of Prehistoric and Dark Age migration (Table 0.1). The survival of a Romanized Dacian population in Transylvania, the base of the modern Romanian state, must be seen in the context of the encircling Carpathians and the availability of easier lines of advance to the north and south-east to Pannonia and the Black Sea coastlands respectively. Furthermore, given the lower latitudes and higher sunshine levels in the south, much of the mountain zone has a significant agricultural potential. The four Balkan countries and the southern half of Hungary have mean July temperatures in excess of 20 degrees centigrade; with the exception of Baragan, Dobrogea, and Moldavia rainfall exceeds 50 cm or 20 inches. The mountains in the southern part of Eastern Europe offer a substantial mineral endowment including fuels (coal, gas, and oil) and ores (ferrous and non-ferrous). The North European Plain has for long been the scene of coal-mining (hard coal in Upper Silesia and lignite in Lusatia and Saxony) and further reserves have recently been discovered, but other minerals are not prominent apart from copper in Lower Silesia and sulphur in Tarnobrzeg.

Equally it cannot be overlooked that the eight nations of Eastern Europe have their own interests arising from ethnic origin, and historical experience as well as the territorial base already referred to. Complexity is only too clearly evident in both cases with a process of migration extending from Prehistory to the resettlement programmes carried out after the Second World War and a turbulent history involving not only the fluctuating relations between nation states but also the conflicting economic and territorial interests of city states and empires. Each state in

Eastern Europe can turn for inspiration to great achievements in the past but none can look back on a history of continuous existence extending even through modern times. In every country therefore there is the reality of a complex inheritance expressed in terms of ethnic minorities, variations in levels of economic development, and acute sensitivity over issues of sovereignty. All too easily it is overlooked that there are specific national characteristics in Eastern Europe, many of them relating to the pre-socialist period and that the modernization process began long before the communists took power after 1945. It is often pointed out that communist regimes are essentially conservative, proudly inheriting much of the bureaucracy and protocol constructed by previous regimes. Equally it could be argued that there is a large measure of continuity in economic development with the accelerated pace of modernization arising not so much from a new ideology of 'socialism' as from an extension of older strategies, making full use of new technology (such as electrification) and mobilizing the population without many of the human rights and financial rewards which westerners would normally expect in return for their labour.

The landscape of Eastern Europe reflects all this very clearly. (3) The growth of settlements and industrial complexes, integrated by a modern infrastructure (including transport, power supply, water, and drainage) is everywhere apparent and the patterns are broadly comparable with those appertaining to other parts of the developed world. But the relatively low levels of personal prosperity can be seen in the emphasis on apartment blocks in the towns, the overwhelming importance of public transport, the limited supply of shopping facilities and other services and the low priority afforded to ecological problems. The complex cultural legacy emerges through settlement patterns, architectural styles and local customs which have been only partially modified since 1945. And the attitudes of people in Eastern Europe are inevitably conditioned by awareness of earlier regimes, although increasingly such awareness arises from the education process rather than from direct experience. Particularly important here is the political geography of the inter-war period when the dismantling of the great empires of Central Europe provided for a degree of national self-expression unprecedented in modern times. While the communist governments characteristically regard their predecessors as economically conservative and socially reactionary, the achievement of independence within specific frontiers is fundamental and few communist leaders have not given the highest priority to national sovereignty. Despite very close economic links with the Soviet Union it is the planning of each individual state as a separate entity that emerges most emphatically from the pages that follow. And the historical process is an essential part of the reality which each national planning process must address.

NOTES

1.   D. Kideckel (ed.) (1983) 'Political rituals and symbolism in socialist Eastern Europe', Anthropological Quarterly 56: 52-104; L. Kolakowski (1982) 'Ideology in Eastern Europe', in Drachkovitch (ed.) 43-53; C.K. Wilber and K.P. Jameson (eds) (1982) Socialist Models of Development, Oxford: Pergamon.

2.   Rugg (1985) 119. See also D.S. Rugg (1970) 'Aspects of change in the landscape of East Central and Southeast Europe', in Hoffman (ed.) 82-126.

3.   I. Berend (1960) 'The historical evolution of Eastern Europe as a region', International Organization 40, 329-46; D.W. Paul (1979) The Cultural Limits of Revolutionary Politics, Boulder, Col: East European Monographs.

Section I

TRADE, TRANSPORT, AND ENERGY

Chapter One

TRADE, TRANSPORT, AND ENERGY: AN INTRODUCTION

Socialist economic development has been based on a political revolution (assuring a monopoly of power for the Communist party and the destruction of the former elites) which has brought sweeping nationalization and rationalization of industry and commerce. But the planning process has been obliged to come to terms with the existing infrastructure of distribution facilities and settlement. Given the great spatial inequalities in the availability of transport, power, and labour, location decision-making has been constrained. But while investment in railways, power stations, and new towns has been undertaken the improvements have been gradual and have generally received lower priority than investment in industry, which has consequently been all the slower to decentralize in the interest of greater equality of access to employment. Because of their superior infrastructure the northern countries have had better opportunities for spreading industrial investments widely (although inherited complexes have continued to dominate the geography). In this section the focus is placed on trade, transport, and energy with particular attention to the co-ordination of the separate national economies through the CMEA or Comecon which has become progressively more effective as an instrument for maintaining a powerful Soviet influence in all East European countries except Albania and Yugoslavia.

TRADE AND THE COUNCIL FOR MUTUAL ECONOMIC ASSISTANCE (COMECON)

Five-year Plans in Eastern Europe involve a large measure of self-sufficiency. (1) But trade with the Soviet Union has always been important and during the early post-war years bilateral agreements effectively disrupted traditional trading patterns while providing the USSR with a group of economically autonomous appendages, each making the greatest possible short-term contribution to the country's economic recovery. (2) It has been estimated that the flow of resources from Eastern Europe to the Soviet Union arising out of reparations and joint company activities during the first post-war decade was roughly 14 billion dollars, similar in magnitude to the transfers from the USA to western Europe under the Marshall Plan. (3) Comecon provided an umbrella of respectability to bilateral deals crudely sustained by coercive measures, and promised some scope for long-term plan co-ordination and specialization on a more equitable basis. The organization was set up in 1949 in order to facilitate contact between member states on economic matters. (4) The decision to act in that year may have been largely due to the launching of the Marshall Plan by the USA which made it particularly important for the Soviet Union to prevent its allies from becoming involved with western schemes of reconstruction. Yugoslavia's defection from the bloc may have been another consideration. But Comecon may also be seen as the economic equivalent to the short-lived Cominform (1947-56) which kept members 'informed' on political matters; it is widely believed that they were both established by Stalin to provide a cover of respectability to the Soviet control of Eastern Europe already secured by the Red Army and by the numerous teams of 'advisers' who penetrated all areas of government. Both organizations therefore 'had small powers and low pretensions because both were formed by Stalin who was paranoid about the possibility that his new satellites should combine against him'. (5) Almost all the East European states quickly joined Comecon: Yugoslavia was naturally barred at first in view of the rift of 1948 but she eventually gained observer status in 1964 as part of a limited rapprochement. Albania on the other hand was excluded in 1961. With several other member states outside Eastern Europe, including Cuba, Laos, North Korea, and Vietnam (and with some collaboration with capitalist countries, demonstrated by the agreement between Comecon and Finland in 1973), the organization accounts for a considerable proportion of Eastern Europe's trade. Surprisingly there was no major economic role for Comecon during Stalin's lifetime because trade between member countries was organized through bilateral contacts (particularly between the USSR and individual East European states). The risk of a growth in East European influence over the USSR that could follow from an increase in Comecon functions had to be set against the positive value of an effective organization as a counterweight to the supranational EEC. The latter stood condemned in Soviet eyes as yet another anomaly of capitalism that was bound to collapse under the accumulated weight of its contradictions but the success of the community has now been accepted as a fact of life and the importance of this market to the communist states has called for recognition, first made explicit in 1977 through the necessity for East European governments (particularly GDR and Poland) to negotiate over fisheries in order to maintain access to community waters in the context of the 200 km fishing limit. In the previous year Romania had stepped out of line and entered into negotiations with the Common Market Commission over textile quotas. However, recognition of the EEC as a supranational organization had not yet been reciprocated. Talks between the community and Comecon date back to 1975 but they have not yet been successful because the former remains unconvinced that Comecon is a genuine equivalent. Instead the EEC has insisted on negotiations with individual communist states. There are doubts as to the effectiveness of Comecon rules in governing economic contacts between the member states and also misgivings over the dominant voice of the Soviet Union. Not unreasonably in view of the immediate post-war realities in Eastern Europe there is an underlying suspicion of Soviet coercion and consequently insistence on direct approaches to individual East European countries is seen as a means of helping them to modify such pressures. It means however that Comecon is stranded in a 'no-win' situation with limited functions as a sign of ineffectiveness and greater powers an unacceptable symbol of a pernicious Soviet influence in Eastern Europe.

Comecon has however made undoubted progress in widening its role. A conference in Budapest in 1955 explored the co-ordination of planned production targets. From these

15

soundings arose a gradual harmonization of Five-Year Plan periods. With the GDR Seven-Year Plan of 1964-70 complementing the 1966-70 Five-Year Plans of other member states (including Albania and Yugoslavia incidentally) all the periods adopted for medium-term plans came into line, although it is not clear which Comecon decision (if any) was responsible. Furthermore various special organizations have been set up under the auspices of Comecon to discharge specific co-ordinating tasks. A spate of such bodies appeared in the early 1960s: the Mir electricity grid dates to 1962, a joint railway wagon pool was created in 1963, and Intermetal (1964) dealt with certain products of ferrous metallurgy which were in short supply. The same year saw the formation of Agromach, with an interest in high-performance agricultural machinery, and OCBI with responsibility for ball bearings, especially special types in short supply. In 1965 Intransmach was formed for the production and marketing of transport equipment for factories and warehouses. Then Intercosmos was set up in 1967 and provided for space research in biology, communications, meteorology, and physics (cosmic resources after 1975), while 1969 brought a co-ordinating body for light chemicals (Interchim) into existence. The organizations were based in the various East European capitals: Budapest (Agromach and Intermetal), East Berlin (Interchim), Prague (the electricity and wagon pool organizations), Sofia (Intransmach), and Warsaw (OCBI). The 1970s gave birth to Intersputnik (1971), concerned with artificial satellite communications systems; Interatominstrument (1972), dealing with nuclear instruments; Interatomenergo (1973) for nuclear power station equipment; Intertextilmach (1973) for textile machinery and Interchimlakno (1974) for chemical fibres.

By way of elaboration it may be noted for example that the Intercosmos scientific programme was drawn up after the initial offer by the Soviet Union to share its cosmic technology for peaceful research on space in 1965. Various satellites have been launched under the programme and these have incorporated some equipment supplied by East European countries, especially Czechoslovakia and Hungary. The greatest successes have been achieved in solar research but East Europeans have participated in manned space flights under a programme drawn up in 1976: it had taken representatives of all Warsaw Pact countries into space by the middle of 1981, beginning with a Czech in 1978 and a

Bulgarian in 1979. A central laboratory for space research was built in Bulgaria in 1969. Through the Intertextilmach organization Bulgaria, Czechoslovakia, Hungary, Poland, and the Soviet Union have come together over the manufacture of automatic shuttleless weaving looms for woollen, silk, linen, and other fabrics. Components from different countries are assembled at the Soviet machine-building factory at Cheboksari which turned out 3,000 machines in 1974 and 4,500 in 1980. Then again there are six working groups to co-ordinate research for the pharmaceutical industry in Comecon. The Soviet Union co-ordinates activity in two groups (antibiotics and cytostatics) as does the GDR (substances for circulatory ailments and nervous disorders), leaving Czechoslovakia with responsibility for additives to human and animal food while Hungary is concerned with steroids.

Comecon: the issue of specialization

Despite the evident interest on the part of the Russians in establishing Comecon as a credible equivalent to the EEC (and the undoubted ability of the Soviet Union to pressurize its neighbours) the communist organization has not been able to achieve supranationality. The reason lies in the simple truth that 'planning governments are far more difficult to integrate than market economies. There is nothing automatic or impersonal about such integration: it is a sustained and direct attack on sovereignty'. (6) In other words it is one thing to lay down clear guidelines for the operation of a free market system across international boundaries, but it is quite another to integrate the planning process which is so fundamental to the communist system. For central planning arises from the monopoly of political and economic power held by each communist government and effective co-ordination of that planning process must inevitably mean that power passes to the higher authority leaving the national government in the menial role of functionary. It must also be conceded that given this critical problem the Soviet Union has not pressed supranationality with any great conviction, which in turn underlines her ambivalence towards Comecon dating back to the organization's inception and her desire for influence over Eastern Europe without however bringing the problems of the area firmly within her own frontiers.

The issue was first brought into sharp focus in 1962 when the more industrialized member states tried to introduce a plan for specialization. The agrarian states would cancel some of their industrialization plans in favour of investment in agriculture and light industry, leaving the more advanced economies to concentrate on heavy industry, particularly the more sophisticated branches. Czecho-slovakia for example had been hit by the falling demand for armaments at the end of the Korean War and sought compensation in East European markets through an end to autarky. In this dispute between advanced and backward economies Khrushchev backed the former. He spoke of the diseconomies of small-scale production and proposed a common planning organ to co-ordinate the economies of the socialist system. It is not clear if this was done pragmatically or out of genuine conviction that the time was ripe for the upgrading of Comecon, although the latter interpretation gains some backing from the appearance of an article in the Soviet academic press dealing with interstate economic complexes (major economic regions transcending national boundaries) which could arise out of a supranational Comecon. Romania generated almost all the opposition since her comprehensive industrialization drive was clearly threatened (and especially the projected iron and steel plant at Galati, seen as the centre-piece). Further, with a long history of conflict with Russia, no Romanian government could be expected to accept a Moscow-centred supranational authority to which the country's own aspirations would be permanently subordinated. The publication of a tactless article on interstate complexes exemplified by the Lower Danube, with Romanian territory combined with adjacent areas of Bulgaria and the Soviet Union, confirmed the Romanians' worst fears that supranationality would result in exploitation of their natural resources for the benefit of others. The Romanian stand was helped by the political backing of China (not a Comecon member state but nevertheless a country capable of exerting great influence against Soviet hegemony) and the prospect of economic support from the west (regarding the supply of industrial plant in the event of any embargo by Comecon countries) in view of the country's good potential for hard currency earnings. Even so the fact that Romania was able to block supranationality single-handedly says much for Russia's ambivalence over Comecon and demonstrates an element of pragmatism in economic matters, in contrast to fundamental issues of Communist party authority and toleration of dissent where there is less scope for compromise. There was even a Russian willingness to assist with the Galati project which was in the end only partially equipped by western plant. Galati is now an established element in the Romanian economy and has even been duplicated by another integrated iron and steel works at Calarasi: it is difficult to imagine that all this was once threatened by Khrushchev's derogatory comments about 'metal eaters' which were meant to cut Romania's industrial pretensions down to size.

The year 1963 saw the end of the first great period of struggle in Comecon. It brought a defeat for the Soviet Union and a victory for Romanian (and East European) nationalism over the vexed question of specialization. Individual member states had the right to develop their economies as they wished. However, the idea of getting an international division of labour was not dropped entirely. After the invasion of Czechoslovakia the Brezhnev leadership tried to promote cohesion and the 'Complex Programme' was launched in 1971. Major integration proposals were also implemented after the 1975 and 1976 meetings with member states introducing special sections in their annual plans to cope with co-operation agreements. The 1984 meeting of Comecon gave approval to the Comprehensive Programme of Scientific/Technical Progress which is to extend to the year 2000. Problems of modernization will now be included in co-ordinated plans, leading to more exchange of technology and the setting up of research teams and design offices. Comecon has therefore struck a balance between nationalism and international co-ordination, but with divisive nationalism remaining a much stronger force than early post-war Soviet leaderships anticipated. (7)

Specialization no longer involves crude discrimination between advanced and backward countries: rather it concerns relatively narrow sub-sectors and individual products, accepting that each country will develop its own chemical, engineering, and metallurgical industries. Another innovation is the adoption of the 'interested party' principle and the lack of any reference to supranational planning organs: member states have a fundamental right not to participate in any aspect of plan co-ordination and any agreed co-ordination is achieved through the co-operation of nation bodies, not by the establishment of supranational authorities. This 'principle of interestedness', with recommendations binding only on those countries officially expressing an interest in the project concerned, provides a satisfactory formula for the Romanians who do not participate in many of the joint projects (often located in the Soviet Union) and do not include any section for integration measures in their national plans.

The impact of Comecon

However in Comecon as a whole progress in the formation of joint enterprises over the past decade has not been outstanding. It is usual to find only two countries involved in most cases, reflecting the administrative problems (complicated accounting) and some governments have been reluctant to get involved because of ideological reservations about the whole concept of international property. A joint investment by the GDR and Poland has been made at Zawiercie where 2,000 Polish workers spin Soviet cotton to produce yarn (12,000 tonnes per annum) required for high quality clothing in the two countries. The German contribution has come mainly through the supply of equipment. (8) In 1976 Bulgaria and Romania formed a joint industrial complex for heavy engineering at Giurgiu and Ruse to produce a variety of complete installations for the domestic needs of the chemical, metallurgical, and mining industries in the two countries and also for export. The joint company Haldex-Ostrava was formed in 1983 and enables the Karvina mining enterprise in Czechoslovakia to use Hungarian technology in the processing of slag heaps to yield coal (up to 10 per cent of the waste material) and by-products to be used for the production of bricks and light concrete aggregates. The processes have been much used in the west and formed the basis of an agreement between Hungary and Poland as long ago as 1959. Several ventures in the Soviet Union are more genuinely multilateral. Thus investments have been made by various countries in the Ulst Ilimsk cellulose factory, part of the territorial production complex based on the Bratsk hydro-electric scheme. Construction work was carried out between 1974 and 1982. East European workers were employed and deliveries of materials included various prefabricated structures for walls and roofs, cement, pipes, cables and electrical, sewerage, and ventilation equipment. Five countries are working together on an extraction and processing complex to handle low-quality ore at Dolinskaya near Krivoi Rog. By the end of the decade it should be producing train loads of ore pellets with a high iron content, which will be relatively efficient to transport. However in recent years joint investment projects have been restricted to energy and transport: the Khmelnitskiy and Konstantinovo nuclear power stations and the 750Kv transmission lines to Poland and the GDR (from Ignalinsk and Khmelnitskiy power stations) and to Romania and Bulgaria (from Konstantinovo). By contrast between 1976 and 1980 there were three energy projects (the Adria and Orenburg pipelines and the Vinnitsa-Albertirsa 750Kw power line) and ten other projects of which six were located in the Soviet Union, three in Cuba and only one (the Varna-Devnya soda plant) in Eastern Europe.

The impact of Comecon is difficult to assess. Structural change has been accelerated through the opportunities for the export of machinery and equipment which brought about a major expansion of the engineering industry. Czechoslovakia has been able to achieve economies of scale in producing various types of equipment for breweries, oil refineries, and textile mills both within the bloc and overseas (in the case of the Basra and Homs refineries). This affects prosperity in different parts of the country since looms and sewing machines come from the Zbrojovka factory at Vsetin and the Elitex factories at Boskovice and Tyniste nad Orlici. Large Soviet orders also bring worthwhile economies of scale to Czechoslovakia's 'town of tailors', Prostejov, which with a long history as a producer of clothing employs some 10,000 workers on suits, supplying around 600 wagon loads to the Soviet Union each year. Comecon agreements have recognized Czechoslovakia's interest in the glass industry: this has medieval origins and gained a high reputation by the nineteenth century with many skilled workers exploiting the raw material and fuel advantages for chandeliers and other lead crystal products as well as the incomparable 'Bijoux de Boheme' with cut glass simulating gem stones and pearls. After initial post-war setbacks through the expulsion of Germans from the Sudetenland there has been new investment. Plate glass produced by the British float method at Retenice meets the needs of the GDR, Hungary, and Poland as well as domestic requirements while mirror glass is made at Teplice. Other products come from a range of factories in Bohemia

21

Table 1.1 Regional distribution of industrial production 1953-83

For each commodity an index is given calculated by dividing the share of each country's output by the share of population. The index relates to the 1983 situation. A value of 100 indicates a 'fair share', and zero indicates no production. The following four figures represent the change in the index over 1980-3, 1974-80, 1965-74, and 1953-65. For total East European production see Figure 6.3 of the companion volume.

Information for the USSR, USA, and Western Europe (FRG, France, Italy, and UK) shows total East European production (excluding Albania) as a percentage of the output for these countries, along with changes in the index over previous periods.

a Metal/Energy Group

	Iron Ore					Aluminium					Copper/Lead/Zinc					Oil Refinery Capacity				
Czechoslovakia	138	-1	+54	-56	-126	73	-3	+5	-14	+39	31	+3	+2	-9	-21	138	+7	-4	-1	+82
GDR	23	-6	-4	-33	-28	62	-1	-11	-81	-34	53	-1	+2	-10	-26	149	+4	+16	+26	+22
Hungary	38	+4	-1	-16	-8	140	+4	+13	-184	-337	23	-1	+2	+17	-23	127	+3	+13	-32	+62
Poland	0	-4	-29	-36	-16	20	-27	-16	-17	+80	145	-9	+13	+17	-55	49	-1	+3	+11	+20
North	32	-2	-3	-43	-39	55	-13	-8	-55	-22	87	-3	+7	+5	-33	98	+2	+5	+5	+34
Bulgaria	259	-10	-7	+79	+158	0					258	+12	-14	-66	+308	156	+1	-20	+69	+91
Romania	93	-16	-4	+5	+42	183	-23	+20	+162	+24	49	+2	+9	+4	+18	126	+1	-1	-77	-213
Yugoslavia	279	+23	+14	+114	+43	212	+66	+1	+39	+69	132	+1	-27	+7	-21	61	-5	-9	+26	+23
South	198	+1	+3	+62	+63	165	+18	+10	+83	+39	118	+3	-11	-8	+53	103	-2	-8	-9	-57

Continuation (USSR, USA, West Europe):

USSR	22	+6	-2	-1	-1	36	-3	-3	+15	+11	56	-3	+9	+9	+2	-6	24	-1	na	na	na
USA	13	+25	-2	+5	+5	16	+4	+1	+4	+4	56	+1	+15	+15	+15	+9	18	+2	+3	+6	+3
West Europe	58	+10	+10	+7	+6	36	+5	+4	+9	+8	59	-2	+3	+4	+4	+13	30	+8	+8	+3	-10

	Energy Output					Electricity Capacity					Electricity Output					Total				
Czechoslovakia	118	-2	-20	-24	+18	135	+5	-14	-32	-4	130	-1	-4	-21	-8	109	+1	+3	-22	-3
GDR	149	+11	+7	-13	+2	145	-6	-3	-29	-14	166	+4	-6	-34	-26	107	+1	+1	-20	-15
Hungary	57		-3	-7		58		-1	-1	-8	64	+3	-3	-7	-12	72	-2	+3	-35	-47
Poland	126	-9	-6	+13	-23	83	-6	-4		-5	90	-4	-2	+6	+1	73	-8	-6	-1	
North	120	-2	-5	-3	-6	102	-4	-6	-14	-10	110	-1	-4	-11	-14	86	-3	-2	-17	-13
Bulgaria	57	+4	+14	-13	+23	169	+51	+1	+37	+49	125	+17	+15	+13	+44	146	+11	-1	+17	+96
Romania	110	+3	+12	+20	+8	89	-3	+9	+31	+10	81	-3	+2	+24	+22	104	-6	+7	+24	-13
Yugoslavia	37	+1	+3	+1	+10	76	-4	+12	+17	+9	75	+2	+7	+15	+21	125	+12		+32	-22
South	71	+3	+9	+6	+12	97	+5	+10	+22	+20	86	+3	+6	+16	+25	120	+4	+3	+26	+22
USSR	24	-1	-7	+9	na	40	+1	+2	+3	+20	36	-1	+2		-10	31	-1	+3	+4	+2
USA	26	+3	-18	-18	na	18	+2	+1	+1	-36	21	-1	+5	+2	+4	24	+2	+1	+1	-1
West Europe	87	-5	-27	+38	na	42	+3	+7	+5		45	+1	+5	+5	+5	51	+5	+1	+10	+4

b Chemical Group

	Caustic Soda					Fertilizer					Plastics					Hydrochloric/Sulphuric Acid				
Czechoslovakia	109	+17	-8	-1	-4	65	-7	+3	-15	+44	204	+4	+45	-8	-87	120	-4	-10	-40	+52
GDR	209	+1	+2	-35	+88	282	-8	+22	-93	-148	197	+22	-21	-38	-22	133	+3	-35	-46	-108
Hungary	86	-6	+29	+11	+10	87	+4	+4	+12	+43	102	+29	+3	+21	+7	80	+10	-13	-6	+2
Poland	54		-19	+7	-5	61	-1	-25	+33	+34	45	-19	-11	+7	+3	77	-8	-13	+19	+7
North	101		-7	-5	-31	111	-4	-8	-11	-23	115	-1	-1	-6	-26	98	-3	-11	-18	-22
Bulgaria	83	-11	+35	+18	+41	113	+3	+9	-8	+88	95	+14	-27	+34	+59	105	+14	+11	+12	+45
Romania	165	-2	+11	+26	+97	123	+12	+33	+38	+39	81	-5		+13	+59	145	+7	+15	+32	+43
Yugoslavia	38	+6	-1	-13	+11	34	+2	-7	+10	+25	68	+1	+17	-2	+26	125	-2	+45	+3	+26
South	98		+10	+9	+51	84	+6	+12	+18	+42	78	+1	+2	+11	+45	103	+4	+17	+28	+37
USSR	93		+14	-15	-7	47	-6	-5	-2	-66	97	-9	+27	-3	+34	43	+1	-14	+1	+7
USA	29	+5	+7	-1	+4	70	+15	+2	+8	+5	35	-1	+22	+2	+6	23	+2	-2	+5	+4
West Europe	46	-1	+15	-12	+3	109	+13	+25	+20	+6	29	+2	+12	+1	+4	51		+10	+6	-5

	Synthetic Fibres					Synthetic Rubber					Tyres					Total				
Czechoslovakia	147	+10	+17	-55	+20	112	+13	-11	-20	+130	115	-6	-27	-1	-122	125	+1	+1	-20	+5
GDR	160	+1	-87	-2	-96	250	+4	-14	-81	-250	176	+19	-20	-14	+42	201	+8	-22	-44	-101
Hungary	35	-1	+14	+8	-7	0					20	-3	-6	-51	+2	59	+6	+5	-1	+14
Poland	74	-16	+18	-18	+8	84	+3	-10	+18	+76	55	-13	-2	-6	-11	64	-9	-9	-9	+16
North	101	-6	-6	-32	-14	113		-11	-17	-23	87	-4	-12	-15	-25	104	-2	-8	-15	-23

Bulgaria	122	+3	na	+78	72	−13	−18	+67	+99	71	+7	−2	+10	+21	94	+2	+12	+30	+36
Romania	132	+11	na	+50	166	−5	+33	+39	+40	93	+9	−3	+5	+44	129	+4	+19	+28	+58
Yugoslavia	57	+9	na	+42	0	−4	+16	+29	na	163	+3	+45	+47	+50	51	+2	−1	+10	+27
South	99	+9	na	+50	81			+23	+40	119	+6	+18	+23	+43	95	+3	−12	+24	+41
USSR	108	−13	na	−12	26		+1	na	na	55	−4	−17	+33	+8	67	−4	−1	+10	−39
USA	22	+3	na	+45	26	+3	+6	+6	+4	18	−4	+10	+7	+3	32	+3	+7	+7	+6
West Europe	44	+2	na	+35	39	+3	+8	−4	na	25		+7	+6		49	+3	+13	+7	+2

c Food Group

	Beer and Wine					Butter and Cheese					Cigarettes					Fish				
Czechoslovakia	173	−2	−21	−37	+4	135	+11	+8	−1	−17	63	+7	−10	−17	−36	11	−13	−51	−25	+70
GDR	162	−3	+24	+2	+18	192	+8	−7	+15	+28	64	+5	+10	−16	−24	131	+5	+6	−14	+22
Hungary	141	+4	+11	+8	+5	66	+8	+20	−8	+4	94	−2	−7	−7		37	−4	+1	+30	−24
Poland	30	−4	−12	+3	−28	117	−10	+2	+15	+15	87	−13	−7	−9		181	−3	−10	+6	−24
North	103	−1	−3	−6		136	+7	+2	+8	−3	78	−3	−3	−12	−11	118	−3	−10	+6	+10
Bulgaria	118	+4	−13	+44		145	+22	−7	−23	+70	395	+30	+1	+126	+132	123	−23	+21	+1	+1
Romania	90	+5	+10	+11	−22	40	−5	−2	−1	+13	61	+5	−7	−10	−15	96	+16	+24	+16	−2
Yugoslavia	94	−2	+7	+16	+7	41	−3	−1	−12	−25	99	−2	+17	+3	+5	32	+5	+4	−12	−32
South	96	+1	+6	+9	+1	57		−3	−10	+7	132	+6	+3	+18	+29	73	+4	+15	−6	−15
USSR	125	−5	+3	−26	−28	68	+6	−5	−3	−1	94	−1	+18	+12	+2	15	+2	−2	+1	+2
USA	52	+2	−3	+5	+12	71	−9	−4	+17	+41	49	−1	+6	+9	+8	36	+1	−14	+22	+18
West Europe	36	+1	+5	+6	+3	47	−3	−1	+3	+15	82	+4	+9	+17	+20	61	+7	+5	+22	+18

Flour / Meat / Sugar / Total

	Flour					Meat					Sugar					Total				
Czechoslovakia	89	+7	-2	-15	-4	114	+2	-6	-11	-20	121	-34	+9	+6	-40	101	-1	-2	-11	-16
GDR	57	+2	-25	+31	+8	140	+4	-4	+17	+19	107	-4	+14	-3	-34	122	+6	-5	+3	+12
Hungary	116	+3	-8	-13	-89	171	+30	+6	-5	+85	129	-5	+55	-46	+15	108	+2	+12	-12	+6
Poland	99	-1	+33	+4	+28	81	-14	-19	-2	-26	137	+46	-44	+4	-10	105	-7	-7	+6	-11
North	91	+2	+8	+5		112	-2	-10		+3	127	+13	-7	-4	-17	109	+2	-3	+6	-3
Bulgaria	152	-3	-31	-16	-93	90	+10	+18	-10	-6	29	-19	-5	-24	-39	150	+3	-2	-2	+27
Romania	102	-12	-7	-12	+47	84	-3	+10	+16	-3	62	-14	-6	+18	+41	76	-1	+3	+3	+8
Yugoslavia	111	+7	-0	-8	-7	79	+5	+20	-7	-2	73	-23	+34	+10	+17	76	-2	+10	-2	-5
South	114	-3	-12	-8	-1	83	+3	+15	+2	-3	61	-19	+11	+8	+31	88	-1	+5	+2	+7
USSR	30	-12	+1	-2	+4	67	-6	+8	+1	+2	65	+21	+13	+7	-48	66	-2	+6	-1	-9
USA	90	-12	-11	+12	+27	51	-2	+4	+11	+9	109	+17	+2	-29	-47	64	+4	-4	+7	+10
West Europe	93	+5	+12	+11	+20	69	-1	-3	+13	+10	63	+17	-17	-7	-9	64	+4	+1	+9	+11

d Other Industries Group

	Cars					Cement					Cloth and Yarn					Paper				
Czechoslovakia	171	+28	-76	-42	-1	129	+9	-6	-12	-11	136	+12	-5	-11	-26	162	+4	+6	-18	-37
GDR	167	+40	-48	-122	-59	134	+5	+5	+1	+11	120	+6	+6	+10	+21	154	+3	-14	-22	+8
Hungary			+37	+42	+10	76		+8	+8	-12	78	+1	-15	-27	-7	92	+8	+9	+16	+2
Poland	108	-10		-18	-28	83	-7	-13	-1	+1	77	-14	-6	-5	-21	68	-5	-12	-4	-16
North	118	+9	-6			102		-5	+1	-1	98	-3	-4	-6	-8	107	-1	-7	-9	-14

Bulgaria	25	+5	-11	+31		119	+14	+13	-10	+34	151	+17	+3	+8	+31	100	+9	+8	+30	+17
Romania	58	-43	+40	+51	+10	108	-6	+4	+13	+1	117	+2	+13	+33	-10	72	-4	+5	+19	
Yugoslavia	108	+8	-4	+20	+84	79	+3	+11	+11	-9	72	+3	+1	-12	+32	102	+2	+17	+4	+55
South	74	-13	+13	+35	+39	98	+1	+7	+8	+3	104	+5	+6	+10	+14	89	+1	+10	+15	+25
USSR	69	-14	+24	-62	+92	55	-6	+8	+5	-34	55	-6	+2		+10	73	-1	+9	-10	-22
USA	13	-4	+8	+6	+3	111		+28	+29	+25	114	-1	+24	+50	+16	11		+1	+1	-3
West Europe	10	-2	+5	+4	+2	66	+3	+13	+13	+7	106	+4	+21	+24	+28	31	+1	+4	+1	+4

	Radio and Television					Roundwood/Sawnwood					Steel					Total				
Czechoslovakia	87	+23	-12	-23	-11	165	-3	+22	+14	-23	210	-8	-9	-13	-42	151	+9	-12	-16	-22
GDR	181	+45	-26	-31	-116	80	-1	+18	+9	-32	94	-2	+7	-11	+1	133	+14	-7	-24	-24
Hungary	78	-20	-8	-14	-1	67	-6	+11	+12	+19	76	+3	-8	-17	-37	67	-2		-6	-5
Poland	134	-20	+27	+43	+10	90	-4	-9	+8	-1	108	+7	-5	-8	+9	95	-8	+3	+11	-1
North	127	+2	+3	+5	-35	99	-4	+6	+11	-9	120	+1	-3	-11	-12	110	+1	-2	-5	-15
Bulgaria	35	+10	-23	-13	+39	73		-16	-53	+81	61	-10	-9	+34	+26	81	+6	-5	+4	+33
Romania	75	-22	+4	+38	+45	120	+8	-16	-22	-16	123	-3	+23	+32	+35	96	-10	+10	+23	+10
Yugoslavia	55	+9	-4	-45	+78	92	+3	+1	-14	-1	21	+1	-13	-3	+13	76	+4	+1	-5	+36
South	60	-3	-4	-6	+58	101	+5	-9	-16	+13	70	-2	+5	+18	+23	85	-1	+4	+9	+25
USSR	41	-13	+9	-10	-30	26		+3	+2	-10	43	+4	+1	+5	-1	52	-5	+8	-10	+1
USA	33	-15	+16	+17	+8	29	-1	-1	+3	-2	62	+3	+20	+13	+13	53	-3	+14	+17	+9
West Europe	42	-9	+20	+1	+8	121	+14		+2	-20	60	+3	+16	+10	+1	62	+2	+11	+8	+4

Source: United Nations yearbooks.

including Hodonin (glass fibres), Kemenicky Senov (illumination devices), Litomysl (glass silk for electro-insulation devices), Novy Bor (decorative glass and glass tableware), Svetka nad Sazavou (crystal glass) and Zelezny Brod (costume jewellery). In addition glass balls and other decorations are still made in workshops in the Giant Mountains. New investment has also taken the glass industry into other parts of the country including Bratislava and Trnava in Slovakia (technical glass and glass fibres respectively) and Valasske Mezirici in Moravia (lighting glass, television tubes, thermoflasks, etc).

It is also arguable that Comecon has contributed to the strengthening of the economies of the poorer member countries. Certainly when output is related to population it can be seen that there is a trend towards greater equality. Table 1.1 presents the results of calculations made into the changing distribution of total East European production of various commodities for which information is given in UN statistical yearbooks. Production details for Eastern Europe as a whole (apart from Albania) are given in the companion volume, the purpose here being to pinpoint changes in the relative importance of individual countries. Share of production is measured against the share of population with parity assigned a value of 100. Figures for 1983 are stated along with changes over the periods 1980-3, 1974-80, 1965-1974, and 1953-65. The increased role of the Balkan countries comes out consistently: from 65 in 1953 to 120 in 1983 in the energy/metallurgy group, from 48 to 85 in the general manufactures group, from 75 to 88 in the food processing group, and from 15 to 95 in the chemical group. Eastern Europe's combined weighting can also be gauged against the Soviet Union, the USA, and the four major industrial states of the EEC. Space does not permit a full interpretation of all the figures but the material may be useful for reference over the present discussion and others presented elsewhere in this book.

The chemical industry

In the chemical industry there has been a transformation everywhere through the switch from a coal to a hydrocarbon base (with much of the raw material supplied by the Soviet Union). Inherited coal-based complexes have been rebuilt and the ease of transport of intermediates (like ethylene and propylene) by pipeline or rail enables downstream processing for fertilizers, fibres, plastics, and rubber to be widely dispersed. Oil pipelines reached both Leuna in the GDR and Zaluzi in Czechoslovakia in the late 1960s.The GDR industry which once embraced a small knot of locations (Bitterfeld, Schkopau, Wittenberg, and Wolfen) now covers additional sites like Bohlen, Premnitz, and Schwarza. Meanwhile the Czech industry embraces Lovosice, Neratovice, and Usti nad Labem. However a feature of Comecon integration is the exchange of intermediates and some specialization in the production of commodities for which demand in any one country is limited. Intermediates have been transferred by pipeline between Bohlen and Zaluzi. In the first instance Bohlen supplied ethylene and propylene (by pipeline and railway respectively) to the polyethylene and polypropylene units at Zaluzi, but the commissioning of new capacities at Zaluzi has since allowed the flow to be reversed. Included in the integrated complex is the works at Neratovice, which produces PVC, and Zeitz in the GDR, which supplies feedstocks to the petrochemical plant at Bohlen.

Hungary's Leninvaros complex is also involved in exchanges. Leninvaros is a major chemical complex which has developed on the banks of the Tisza, including a power station (Tiszapalkonya), oil refinery (Tiszazederkeny, processing crude oil received by pipeline from the Soviet Union), and various chemical factories. It now occupies a focal position in the Hungarian chemical industry with intermediates supplied to the Borsod factory at Kazincbarcika (near Miskolc) and others at Pet and Szolnok. There is a 335 km pipeline between Leninvaros and Kallus in the Soviet Union (opened in 1975) so that Hungarian ethylene (and propylene in tank wagons) can supply the Soviet vinyl chloride factory. Just over half the total ethylene production of 250,000 tonnes per annum and four-fifths of the total propylene output of 100,000 tpa is sent to Kallus, enabling Hungary to gain economies of scale in these two operations. Furthermore Hungary will receive from the USSR various products required in quantities which are too small to justify production in Hungary: acrylonitrile, ethylene glycol, and polystyrene. Poland's petrochemistry focuses on two complexes, one at Plock linked with downstream processing in the north of the country (Bydgoszcz, Olsztyn, Poznan, Pulawy, and Torun) and Blachownia Slaska with a complementary role in the south covering Brzeg Dolny, Jelenia Gora, Kedzierzyn, and

Oswiecim. Crude oil from the Soviet Union is the essential underpinning.

In the Balkans the Comecon dimension is less important. Yugoslavia's chemical industry focuses on the Industrya Nafte complex on the island of Krk near Rijeka, built with American assistance and using Middle East crude; while Albania and Romania have their domestic hydrocarbon resources and have built up capacity at Fier and Pitesti/Ploiesti respectively. (9) But Bulgaria's chemical industry rests very heavily on Soviet expertise and raw materials. The communist government inherited a small oil refinery on the Danube at Ruse and initially went ahead with a coal-based chemical complex at Dimitrovgrad in 1951 (spawning a superphosphate factory in 1957) and a soda factory using Soviet raw materials at Varna-Devnya in 1954. Attempts were made to develop domestic reserves of hydrocarbons, with fertilizer production at Vratsa (built 1964-8) to use the Chiren gas and oil refining at Dolni Dabnik near Pleven. However these resources have been insufficient to permit economies of scale and the chemical industry has been rebuilt on the basis of Soviet crude oil shipments to the Burgas refinery (where petrochemical production began in 1963, extending to synthetic rubber and polystyrene by 1970) and Soviet gas piped across Romania to a circular distribution system in northern Bulgaria (1974) supplying a range of chemical factories: some existing (like Vratsa) and others newly built. Heavy concentration of production is thus being avoided as Bulgaria manufactures a limited range of commodities and trades with the Soviet Union to obtain the goods for which home demand is too small to justify plant in Bulgaria. An ethylene pipeline from Burgas feeds the Devnya complex which went into the production of chlorine in 1964, mineral fertilizers and calcinated soda by 1974 and PVC in 1980. Further investments made during the 1980s have boosted production of phosphoric acid and polyvinyl acetate, while the industrial complex has been broadened by the growth of cement and sugar factories and the construction of the new Devnya harbour. But intermediates from Devnya (and Burgas) also supply a chain of downstream processors like Vidin which produces polyamides using polyethylene from Burgas and PVC from Devnya. Certain highly specialized plastic goods are produced at Ruse including flooring materials and polyurethane artificial leather used in the production of special clothing for the building and mining industries. Synthetic fibres have been produced by Dimiter Dimov at Yambol and Sviloza at Svishtov since 1968 and 1973 respectively, pharmaceuticals at Razgrad and Troyan (making maximum use of Bulgarian medicinal herbs and technologies), and fibreglass at Knezha. Other locations include Pazardjik, Stara Zagora, and Yambol. Thus the Bulgarians can claim that Comecon has meant the growth of a major industry with a distribution which extends not just to the ports but to many other districts once predominantly agrarian.

Prospects

However bilateral arrangements prevent the full use of each country's potential to participate in an international division of labour. They capture resources which might otherwise be allocated to projects to increase exports outside Comecon and generate hard currency with which to finance imports from the west and to secure scarce goods from Comecon members. They may also maintain some disparities between the more developed and less developed Comecon states, reinforcing the position of the latter as raw material producers. Unease is deepened by problems of pricing and convertibility. The currencies of Comecon countries are not convertible, so supply and demand play no role in fixing their international value. Instead currencies are arbitrarily fixed by the authorities with the result that the valuation of the commodities traded is in turn arbitrary. Consequently a particular package of exports may attract more goods in exchange from one socialist country than from another: it is all a function of the bargaining strengths of the parties negotiating bilateral trade agreements. In the absence of a solution to the problem Comecon countries have resorted increasingly to using world market prices, although identifying world prices is much more difficult for manufactured goods than for raw materials. The inadequacy of the Comecon clearing system has led to the settling of accounts in dollars or some other western currency. This has been particularly noticeable in Hungary's dealings with her partners.

Comprehensive reform of Comecon trading procedures has always been opposed by the Soviet Union. A system of 'transferable roubles' allowed for some tidying up of accounts where surpluses with some countries could be balanced by deficits with others, but with rising oil prices during the 1970s virtually all the East European members of Comecon have been running up deficits in Moscow. These subsidies to Eastern Europe might well persuade the Soviet Union to contemplate substantial reforms over pricing and convertibility. But this is a small price to pay for greater bloc cohesion given the dilemma faced by the East European allies. The Soviet Union has been pleased to use its abundant fuel and raw material resources as an instrument of economic integration which helps to maintain its hegemony over Eastern Europe as compliant behaviour by its allies may be rewarded by economic favours. Retreat under the Comecon umbrella is therefore an instinctive response by Eastern Europe to economic crisis, but there may be profound economic constraints over the longer term if many industrial requirements can be met only by western suppliers and expansion of capacities through switches to Comecon technology may affect export opportunities. Dependence on Moscow will be minimized only by increases in economic efficiency and competitiveness but this is possible only with far-reaching reform of the central planning system. (10)

TRANSPORT

Post-war demand for transport services has risen sharply on account of the emphasis on the expansion of heavy industry, often involving imported raw materials. Most prominent in transport flows in Eastern Europe are the commodity transfers from the Soviet Union. In 1975 these transfers were headed by oil and oil products totalling 67.7 million tonnes (52.0 per cent of all Soviet exports of these commodities), 38.1 million tonnes of iron ore (87.2 per cent), and 16.4 million tonnes of coal (62.7 per cent). Also very significant were the exports of rolled iron/steel (5.8 million tonnes/89.6 per cent), pig iron (4.3/90.8), apatites (4.1/70.0), potassium salts (3.9/65.8), and coke (3.1/73.6). Reverse commodity flows are dominated by Polish coal, amounting to some 10 million tonnes annually. Overseas trade has meant a big increase in the importance of shipping while the national airlines operate many international services for commercial travellers, diplomats, and tourists. However, it is important to stress the autarkic element in all transport systems. Relatively few international links are provided and both railways and roads have often been cut at the frontiers.

Within each state the planners have come to base their thinking on urban agglomerations and this gives the greatest importance to the transport corridors connecting them. Railways have been developed with additional tracks and with electrification in many cases while surfaced roads (sometimes motorways) and air services are also provided. At the same time notions of equality of access to jobs in manufacturing have led to a dispersal of light industries as the extension of the electricity grid has made power available in virtually all settlements. Existing centres of heavy industry inevitably retain their importance but since 1960 many backward areas have received factory industries for the first time and the regional centres have grown very rapidly. So greater transport capacities in the backward areas are therefore required. And finally it should be remembered that strategic considerations make it imperative that all parts of each country are immediately accessible from the capital.

Transport planning: the central role of the railway

Central planning in Eastern Europe inevitably gives a high priority to transport to ensure that the various demands can be met. But transport is not considered to be 'productive' in the material sense and the planners therefore aim at a 'unified transport system', with the fullest co-ordination between the different transport media. There is very little duplication, seen as actually or potentially wasteful, so that any lengthy journey, for goods or passengers, usually involves a media mix. The railway (see Figure 1.1) is fundamental for both goods and passengers but short journeys of less than 50 km are increasingly being made by road while very long journeys are often accomplished by air, by passengers at least (Table 1.2). Long freight hauls still use rail, but shipping (by sea, river, or canal) is an option in some cases, while bulk transfers of oil and natural gas are frequently made by pipeline. In Poland four-fifths of all road freight traffic involves distances of less than 30 km while a similar proportion of rail traffic involves distances greater than 50 km. Over distances greater than 75 km the cost of rail transport is considered to be between four and six times cheaper than road (there is a 'substitution zone' for both modes between 30 and 75 km). The mean distance travelled by passengers in 1981 was 21 km for road and 48 km for

Figure 1.1 The East European railway network

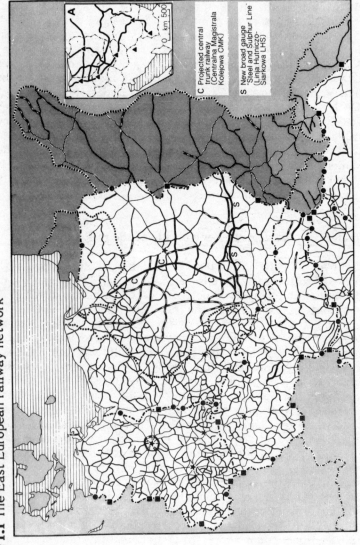

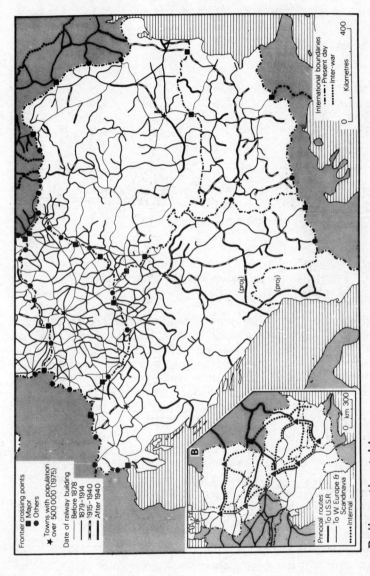

Source: Railway timetables.

Table 1.2 Prime hauliers of freight traffic among Comecon member countries 1950-90 (a)

Haulier	Percentage share (b)				
	1950	1960	1970	1980	1990
Railway	89.3	87.9	66.1	49.4	47.4
Sea	7.1	7.5	12.7	19.4	20.4
River	2.6	4.5	4.6	4.5	4.5
Pipeline	*	*	16.4	26.4	27.4
Road	*	0.1	0.2	0.3	0.4

Source: R.N. North (1984) The Soviet Transport Interface with Eastern and Western Europe, Washington, DC: Wharton Econometric Forecasting Associates.

Notes:
(a) Includes Cuba and Vietnam after joining Comecon.
(b) Takes the middle value when a range is given.
*   Less than 0.1.

railway but the difference is tending to increase as more suburban journeys are made by bus. Whereas the number of rail journeys has been fairly stable since the late 1960s the number of bus passengers has increased rapidly with continuing urban expansion and relatively low fares combined with a severe fuel shortage for private cars. (11)

All this does not necessarily lead to greater efficiency. Between 1980 and 1985 the GDR envisaged a sharp reduction in the role of road transport for freight (from 26.4 to 15.0 per cent) while railways intensified their activity (70.9 to 82.0 per cent) and inland waterways made modest progress (2.7 to 3.0 per cent). However this meant that enterprises could not exercise the choice that had been available in the past and had to pool road vehicles which have previously been under their own control. (12) Conversely in Poland enterprises geared to rail transport sometimes responded to the minimum distance regulations for rail transfers by deliberately using distant suppliers because the transport of supplies available nearby would have to be by road. Transport is heavily subsidized and costs to the manufacturer therefore comprise too small a share of total production costs to stimulate rationalization. Again a railway business which does not have to make a profit can claim bottlenecks as a lever to achieve further investment to meet plan targets. Strong political pressures can therefore accomplish remarkable results, as in 1984 when the Polish railways moved 43.1 million tonnes of coal for export although it was claimed that 38.0 was the best that could be envisaged.

Further problems arise from inadequate consideration of real transport costs in decisions on industrial location and on optimum factory size. In Poland there is not only demand for bulk transport between Upper Silesia, Warsaw, and the Baltic but also an additional need to support a particularly widespread factory system shaped by the plans of the 1950s which abandoned the idea of major concentrations of industry at Pila, Sandomierz, and Wizna in preference for a strategy of nationwide industrialization. Poland therefore incurs high transport costs in relation to its total manufacturing effort. (13) The economy is highly transport-intensive and the increase in 'transport-absorbtiveness' which has continued steadily through the 1960s and 1970s involves heavy financial burdens. To some extent these costs arise from the nature of resources and their location but the problem has been exacerbated by too much specialization, which means that some bulk manufacturers have to distribute their goods (such as different grades of cement) over large areas and assembly firms have to draw components from an excessive number of suppliers: the Warsaw car factory co-operates with 600 suppliers and the ship-building industry deals with 7,000 subcontractors located throughout the country. Closely related to this problem is over-concentration, arising out of the industrial 'gigantomania' of the 1970s. This means lengthy rail hauls and, in the case of the food industry, some deterioration in quality between factory and consumer. Irrational location also imposes burdens on the transport system: some food-processing units like the sugar factories of Lapy and Ropczyce are not situated close to their raw material supply areas while chemical works like the Olsztyn tyre factory are remote from both raw material suppliers and markets.

Some further complications arise because certain transport systems rest heavily on nineteenth-century foundations. The development of airlines and motorways is fairly recent but substantial railway and inland waterway networks were inherited by all post-war governments except the Albanian regime. The territorial upheavals in Eastern Europe in 1918 posed great problems for the emerging successor states. The new state of Poland inherited parts of the Austrian, German, and Russian railway systems. She had to adapt the latter's broad-gauge lines to accommodate standard-gauge traffic and construct new lines in the centre between Warsaw and Poznan and Warsaw and Krakow as well as the coal line from Silesia through the 'corridor' to the Baltic. Elsewhere large developments were required to link the Czech Lands more effectively with Slovakia, Transylvania with the Danube provinces of Romania, and Serbia with Yugoslavia's Croatian and Dalmatian regions. Very often the completion of these programmes was delayed until after 1945 when further territorial changes had to be taken into account. The historical legacy is still very considerable; while the 'Sudbahn' and the Krakow-Bucharest routes now run rather illogically through corners of Yugoslavia and the Soviet Union respectively, former frontier cities like Lodz in Poland and Brasov in Romania still lack the importance for rail traffic that their central positions would now suggest.

Moreover, despite the importance of rail services the railway systems of all the East European countries are still incompatible with the Soviet system, on account of differences in rail gauge, loading gauge, axle weights, train lengths, and train speeds. A great deal of freight traffic is trans-shipped at some fifteen frontier crossing points, the most important of which are with Poland (Brest and Mostiska) and Czechoslovakia/Hungary (Chop). In some cases, however, it is possible to change to bogies of wagons (using equipment developed in the GDR): this applies to the frozen traffic between Hungary and the Soviet Union (since 1983) and vehicles using the Ilichevsk-Varna train ferry. In the case of passenger trains the bogies of restaurant and sleeping cars may be changed. There has been some progress in extending the Soviet rail gauge (1,524 mm in contrast to the usual 1,435 mm) to Galati and Kosice. But while schemes have been mooted for more ambitious projects extending the Soviet gauge into Hungary, Poland, and even the GDR, the only one recently implemented is the Polish LHS running from the frontier at Vladimir Volynskii to Slawkov near Dabrowa Gornicza, the location of the Katowice metallurgical plant. Standard-gauge lines cross the Soviet frontier at Brest, Reni, and (planned) Uzhgorod but only for short distances to facilitate rail-water trans-shipment.

Road and waterway transport

Transport networks still vary in density between regions as well as countries (see Figure 1.2). In Poland the region (voivodship) with the densest road system has three times as many roads as the poorest region (the ratio is 4:1 if modernized roads are considered) but for railways the extremes are far greater at 9:1 (Lomza has 2.7 km of railway per hundred square kilometres of territory, and Katowice 24.1). (14) Current demand has a bearing on this but there is an important historical dimension arising from the greater emphasis on railway construction in the former Prussian territories before 1918. However there is a tendency to compensate for limited railway services by better road services. Bus services have developed so as to complement rail services and so they operate principally between towns and cities lacking a direct or speedy rail connection. In Poland the density of such services is very high in the east (former Russian-administered territory where the railway network remains sparse), whereas in the former German and Austrian lands railways are much more important. In Romania bus services operate on the several Carpathian routes which, on account of the old frontier with Austria-Hungary along the crest of the mountains, have never been followed by railways; the Eastern Carpathians demonstrate the point especially well with one bus service operating from the railhead at Vatra Dornei along the Bistrita Valley to Bacau and thence across Moldavia to Iasi.

Roads play an important part in most countries, especially in the industrial regions of Bohemia, Halle-Leipzig, and Silesia. Some new roads have been built, notably the Dalmation Highway in Yugoslavia and sections of motorway on the principal routes out of most capital cities in Eastern Europe. A motorway now connects Prague with Bratislava and Sofia with the Black Sea ports but Poland has been forced to postpone a motorway programme with only the dual carriageway between Warsaw and Katowice complete (1973-6). So there is little evidence yet of any international motorway scheme. Indeed, the reverse is more demonstrable since the inter-war German autobahns from Berlin towards Stettin (now Szczecin) and Breslau (now

Figure 1.2 The East European main road system

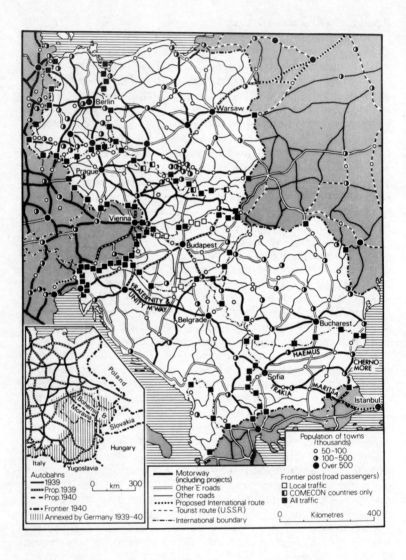

Source: National atlases.

Wroclaw) now peter out short of the present frontier with Poland and the former autobahn in East Prussia is abandoned. Most improvements are still concerned with the upgrading of existing roads: surfacing, with some widening and realignment. However the international dimension is becoming stronger. Bulgaria began with a motorway ring project, building the Trakia and Haemus motorways to connect Sofia with Burgas and Varna respectively (also the Cherno More motorway to run from Burgas to Varna) but the Maritsa motorway now runs to the Turkish frontier and the motorways to Varna connect with the ferry service opened in 1978 to Ilichevsk in the Soviet Union. Bridge-building across the major rivers has been a feature of post-war road transport development, notably along the Danube in the vicinity of major cities like Bratislava. At Budapest the Lanchid bridge of 1842-9 still stands, supplemented by others such as the Margit (Margaret) and Szabadsag (Liberty) bridges which accommodate the Inner Boulevard as it describes a circuit around the historic core.

International road transport of freight is limited but there are regular services from Moscow to Berlin and Sofia; and further development of long-distance lorry transport seems likely in view of the decision to upgrade certain roads during the late 1980s: principally Gdansk-Vidin, Moscow-Berlin, Moscow-Sofia, and Rostock-Constanta. Most trade between Eastern and Western Europe is handled by sea or rail but there is a substantial amount of road transport carrying perishable commodities. Moreover several road services between the Soviet Union and Western Europe cross Eastern Europe in transit (for example Moscow to Copenhagen, Hamburg, Paris and Rotterdam, and Togliatti to Turin). One highly specialized form of road transport concerns the deployment of lorries fitted with electronic equipment in Scandinavia. These vehicles, carrying the international TIR carnet which allows them to cross frontiers without routine customs examination, travel regularly from the Soviet Union through Finland, Sweden, and Denmark and then cross by ferry to the GDR. Lorry fleets for international road transport are now quite large. The Hungarian firm Hungarocamion possesses some 800 lorries which carry normal and refrigerated goods to all parts of Europe and the Middle East. The fleet handles Hungarian foreign trade requirements and also carries foreign goods especially in parts of the Middle East where railway and port transit facilities are overstretched, and

41

where a quarter of the fleet has been operating continuously. The company's operations have been enlarged by the formation of affiliated firms in Austria, Italy, and Kuwait. Czechoslovakia, another land-locked country, also has important road transport (CSAD) enterprises co-ordinated by the Cesmad organization. A fleet of some 2,000 lorries operates over Europe and the Middle East handling a significant part of the country's foreign trade. It maintains the largest trucking trans-shipment centre in Eastern Europe (at Prague-Hostivar) with 10,500 square metres of covered space and is represented abroad through a chain of offices in the socialist countries (Budapest, Moscow, and Rijeka) and in Western Europe (Hamburg, Koln, Rotterdam, and Vienna). In 1973 a Comecon agreement provided for assistance to motor lorries and their drivers in the event of emergency and Czechoslovakia now has a transit service which can help vehicles from non-socialist countries. Repair for Volvo vehicles is already good because Czechoslovak road transport enterprises own many vehicles of this type, but facilities for the repair of Daimler-Benz and Ford trucks are available at Brno and for Iveco at Horovice and Cesky Tesin.

There are some reservations in the west over the activities of the East European road hauliers because they are favoured by trade contracts which usually specify transport by the East European partner. (15) Western transport firms gain only a very small share of the business and hence de facto transport monopolies by eastern state-controlled trading and transport organizations arise. The situation with road transport is that lorries from Eastern Europe are free to seek out return loads in western countries, while protracted customs formalities face those western operators whose vehicles do manage to participate in trade with socialist countries: this is not an encouraging precedent. The problem extends to the waterways because of the East European intention to exploit the free navigation on the Rhine, while failing to ensure similar freedom for western boatmen working (for example) through the GDR to Szczecin has created some demands for revision of the Mannheim Convention of 1868. The prospect of many more vessels from the socialist countries on the canals of the FRG and the consequent danger of more unfair competition does nothing to increase enthusiasm for the completion of the Rhine-Main-Danube Canal. The principal waterways of Eastern Europe are the Danube and Elbe, each of which have links, by canal and/or coastal shipping, with both Western Europe and the Soviet Union. Passenger services by river boat have some local importance and some international tourist cruises operate on the Danube but the main function of the waterways is a modest contribution to freight handling.

International road passenger transport and the growth of 'shopping tourism'

As regards passenger transport by road there are some international bus services operating on a scheduled basis in Eastern Europe, in addition to the many summer tourist services to the Black Sea coast and other resorts. But fuel shortages have constrained development of the network in the 1980s. After laying down complications in 1979, by insisting that all citizens from socialist countries pay for petrol purchased in Romania in hard currency, Romania had killed off all international bus traffic by 1983 by withdrawing her own services and denying facilities to erstwhile partners. Services between Budapest and the Romanian towns of Alba Iulia and Cluj-Napoca had been well used by members of Romania's Hungarian community since the bus timings were often more convenient than those offered by international trains for settlements in western Romania and eastern Hungary. However the private car is becoming more prominent. Car ownership for long remained a remote possibility and households which could command the savings necessary to secure purchase were discouraged by the very limited supply of vehicles, a poor infrastructure (road system, fuel supply, repair service, etc) and a hostile propaganda which related car ownership with a petit-bourgeois mentality. But most countries now have a car assembly industry, in addition to bus, lorry and tractor capacities. The result is that a considerable amount of road traffic is geared to recreation, supported by a growing number of motels and tourist facilities. Levels of car ownership lag far behind those of North America and Western Europe but there are significant variations within the region. In 1973 21 per cent of households in Czechoslovakia and the GDR owned a car compared with 14 per cent in Yugoslavia, 12 per cent in Bulgaria and Hungary, 8 per cent in Poland, and only 4 per cent in Romania. Difficulties for motorists still arise through poor garaging facilities, high petrol costs, and even some restrictions on the use of cars designed to reduce demand for fuel.

Motoring by western visitors has become very prominent now that virtually all main roads are modernized. There is also a considerable international movement between the socialist countries although it is constrained by bureaucratic controls. As already noted, travelling across well-guarded and in some cases fortified borders is restricted to a small number of crossing points. Movement is also heavily constrained by elaborate formalities over the supply of passports, visas, and foreign currency which make any proposed visit a complicated and expensive business. Only in Yugoslavia is the traveller generally unconstrained by official discouragement and even in this case the serious losses of foreign currency led to a new law on travel deposits in 1981. There has, however, been a noticeable slackening in the amount of travel between socialist countries by car (and to some extent by bus and train also) as a result of measures taken to discourage 'shopping tourism'. Relatively straightforward procedures for the acquisition of passports, visas, and foreign currency along with liberal customs regimes allowed people to go abroad in order to buy goods which were either cheaper than those at home or simply not available at all on the domestic market. This practice of exporting shortages was a spontaneous activity which planners found highly frustrating but it was accepted by the authorities for a time as a necessary consequence of 'fraternal co-operation'.

Border shopping raids by Poles crossing into the GDR have been responsible for the collapse of the open-frontier regime for the Oder-Neisse line which dates back to 1972. Arbitrary values for currencies and commodities combined with easy access to foreign currency (facilitated by a black market in Poland) enabled Poles (often working for friends and colleagues) to empty stores of foodstuffs and consumer goods in GDR border towns much to the chagrin of the local inhabitants whose criticisms of 'Einkaufstourismus' began to appear in local newspapers. The Germans however could profit from highly attractive black market exchange rates for their Ostmarks in Poland and therefore extend their vacations or indulge themselves with a higher-than-usual living standard. However the high level of tourist traffic (723,000 per annum in 1970 and 1971 rising to 12.2 million per annum on average between 1972 and 1978, covering the flow in both directions) brought a return to restrictions of increasing severity after 1978. The crisis in Poland in 1980 increased pressure on Czechoslovakia as well as the GDR, leading the authorities to impose strict currency and customs controls and insist on an 'invitation' as a prerequisite for issue of a visa.

In 1981 similar action was taken by Bulgaria and Hungary, responding presumably to food shortages in Romania as well as Poland. Yugoslavia maintains a relatively large number of frontier crossing points and Yugoslavs have developed considerable enthusiasm for 'shopping tourism' in Hungary as well as the country's capitalist neighbours. Yugoslavs like to visit Pecs in Hungary for shopping purposes while Graz in Austria is a popular destination for discriminating Yugoslavs who wish to buy the best coffee (which includes caffeine-free Franck coffee produced in Yugoslavia and exported to Austria to earn foreign currency!) However, the movement was sharply reduced in 1981 by the reduction in the level of duty-free imports in a bid to reduce the drain on foreign currency reserves and ease pressure for devaluation. Meanwhile shopping tourism is developing among Austrians who like to travel to Hungary. In this case however the Hungarians positively encourage the business which has expanded even further since a branch of Marks & Spencer opened in Sopron (seven days a week).

Co-ordination of production of transport equipment

The 1971 Comecon comprehensive programme has made possible co-operation in the development and production of motor vehicles. Multilateral arrangements have made Czechoslovakia the sole producer of lorries of twelve tonnes and over: lorries produced by Tatra Koprivnice are exported in large numbers to the Soviet Union where they work under rigorous Siberian conditions: the lorries have been invaluable in the Far East in areas like Magadan where land transport revolves almost exclusively around motor lorries. The specialization agreement has meant a doubling of production by Tatra during the 1970s and this has benefited the main plant at Koprivnice and also branch factories at Banovce, Cadca (both in Slovakia), Novy Jicin, and Pribor. Tatra lorries have also been purchased by China and have gained a still higher international profile following successful participation in the Paris-Dakar rally (since 1985).

Collaboration has secured an important stake for the weaker members of Comecon. Bulgaria's motor industry has developed out of repair shops set up with Soviet assistance immediately after the war in Plovdiv and Sofia. A larger works at Shumen began to produce spare parts, some of which were exported to the Soviet Union in 1960 and went into truck assembly (the Russian GAZ-53A vehicle) in 1967. Thanks to Czech assistance the Shumen factory has now become a major producer of axles for the bloc and axles for front-engine trucks are now being made. Also, in 1970 the Lovech factory assimilated Czechoslovak technology for its Madara truck (now Skoda-Madara) and co-operation with LIAZ has resulted in the production of LIAZ-Madara trucks, beginning at Shumen in 1983. A feature of the co-operation is the development of an improved back axle which will be supplied to the LIAZ factory in Czechoslovakia. Indeed with a network of component suppliers and other units dealing with automobile superstructures (Preslav) and trailers (Novi Pazar) only the clutch and motor for the Madara trucks come from abroad. Some Perkins diesel engines are made under licence at the Vassil Kolarov factory in Varna. Bulgaria also produces electric trucks mainly for export at the factories of Dounav (1973) and Septermvri 6 (1963) in Lom and Sofia respectively but also at smaller units at Pleven and Vratsa. Electric hoists are made at Gabrovo and Gorna Oriakhovitsa.

However, integration can lead to problems for enterprises which have to give up certain types of production. The vehicle factory at Csepel in Budapest had to cease production of lorries and tractors on account of Comecon specialization agreements. The enterprise had not succeeded in selling enough of its own engines and finished vehicles to justify the expansion of the works carried out in the early 1970s in connection with emphasis given to the road vehicle industry in the 1971-5 Five-Year Plan. Not only was the investment compromised by developments in Comecon but also by one of Hungary's own enterprises, the Ikarus bus factory chose to use engines manufactured at Gyor, under licence from the MAN company of the FRG, in preference to the local Budapest-Csepel product. Csepel has been forced to retool to produce the modern type of bus chassis and steering units for lorries. The enterprise has also established a joint company, with Volvo of Sweden, for production of the Volvo Lapplander cross-country four-wheel-drive truck. Volvo will supply certain components in return for which there are export opportunities for some of the vehicles produced in Hungary: the South American market is a particularly good prospect. Hungary also manufactures the American Steiger tractor under licence in Budapest while production of small tractors and other equipment suitable for use on small-holdings is now being produced at Veszprem under licence from various firms in the FRG, France, and Italy. The leading Comecon producers in Eastern Europe are however in Czechoslovakia. Heavy tractors are made at Martin in co-operation with the Polish firm Ursus and lighter vehicles (including a type specially designed for use in the mountains) are built by Zetor in Brno, with some export to western markets.

Czechoslovakia is the leading East European producer of motor cycles. Diversification at the Janecek armaments factory (founded at Nusle, Prague, in 1922) led to the production of the first motor cycle in 1929 in collaboration with the German Wanderer company (hence the name Jawa). Production continued after the war but relocation in 1963 to Tynec nad Sazavou, 35 kilometres from the capital, was marked by a failure to improve the design at a time when Japanese manufacturers were making rapid progress (helped to some extent by the licence taken out by Honda in respect of the Jawa automatic clutch). Production continues but exports to the west are now limited. However the much more important car industry of Czechoslovakia has developed out of motor cycle production at the turn of the century. V. Laurin diversified from bicycles into motor cycles in 1898 and then cars in 1905. This laid the basis of the Skoda car enterprise because Laurin & Klement merged with Skoda in 1925. Unfortunately the requirements of the Soviet Union in the 1950s were not conducive to an expansion of car production and the opportunity of gaining an early foothold in the western market was lost. The factory at Mlada Boleslav was however modernized and enlarged in the 1960s and exports to Western Europe increased at this time. Skoda Great Britain was set up in 1965 and the company now has a large service and storage centre at Kings Lynn, where all cars undergo rigorous pre-delivery service and are equipped with optional extras to meet the needs of individual customers who can place orders through a network of 250 dealers. The company has gained a reputation for reliable and well-constructed family saloons but can now offer a high-performance convertible as well. Skoda cars are also exported for assembly abroad, notably at the Coopesa assembly plant at San Jose in Costa Rica. Deals with Indonesia, Ireland, and New Zealand have also been reported. Skoda had to cope with a very serious fire in 1969, considered the worst accident to affect Czech industry in the post-war period, but the firm went on to produce new models during the 1970s and a significant research and development effort has been maintained. A car-testing track has been built at Most on land previously used for open-cast lignite working. Some success has been achieved in international races and rallies.

However car production in Eastern Europe generally was distinctly modest in the 1960s and Comecon integration has tended to favour the relatively strong Soviet enterprises which met much of the limited demand in the early post-war years. It is true that car headlights manufactured by the Autopal enterprise at Novy Jicin in Czechoslovakia, one of the many component suppliers to Skoda, are exported to other socialist countries and especially the Soviet Union, whose Togliatti works have placed a major order. Also some useful collaboration has developed with Bulgaria since the Moskvich 408-type car was first assembled at the Balkan works in Lovech in 1966. This enterprise continues but with Bulgaria now providing some components (notably the batteries which are made at Tolbukhin). The Zhiguli is also assembled in Bulgaria with air filters, dynamos, and starter motors, as well as batteries, manufactured within the country. Most of the parts are made at a factory in Sliven which opened in 1970 and which supplies some two-thirds of all the dynamos and starter motors required in the Soviet factory (Togliatti). There have been strong rumours that a projected joint initiative by Czechoslovakia, the GDR and Hungary for production of a new car at the Zwickau works has been discouraged by the Soviet Union, fearing competition for the Zhiguli.

An alternative strategy can then be collaboration with western firms for there is a continuing interest in western vehicles. There is always demand from affluent and well-connected families; for the state occasional purchases can assist with negotiations for wider trade contacts and possible installation of vehicle factories in Eastern Europe. Interest in both Czechoslovakia and the GDR has been shown in Volkswagen (whose Golf/Rabbit and Passat models have done well in the GDR) and both states could potentially become Volkswagen subcontractors. Poland made arrangements with Fiat in 1979 for production of a range of cars at

Bielsko-Biala using some Italian parts and other components manufactured in Poland. Fiat had previously entered into a scheme of technological and financial co-operation with the Yugoslav firm Crvena Zastava (based at Kragujevac) whose first passenger cars were built under licence in 1967. Further agreements in 1974 and 1978 increased Fiat's investment and involved co-operation in foreign markets. They also induced a willingness on the part of the Italians to help finance the construction of motorways in Yugoslavia. Crvena Zastava have also been able to work out co-operation deals with firms in Poland and the Soviet Union. Their production increase has justified a new factory built at Zagreb. Meanwhile Volkswagen have been co-operating with TAS in Sarajevo, and Peugeot have invested in a new car factory in Pristina.

The outstanding East European bus producer is the Hungarian firm Ikarus of Szekesfehervar which has assimilated western technology as a means of increasing its penetration of Comecon, including the supply of various components (engines, starters, and gear-changing mechanisms). Ikarus has successfully penetrated the American market, forging a partnership with the Crown Coach Company of Los Angeles, which will allow the assembly of buses with Hungarian frames, bodywork, and fittings. Ikarus have also sold buses to Greece, Turkey, and various Third World countries. Ikarus retain some links with East European technology. Back axles are supplied by the Raba factory at Gyor, which has assimilated the design used by the Czech Karosa enterprise. Some of Raba's production goes to the Karosa factory at Vysoke Myto. Collaboration also exists between Czechoslovakia's LIAZ enterprise at Jablonec and the Jelczany motor works in Poland: LIAZ supply the chassis and the Poles contribute the bodywork according to Czech blueprints. Bulgaria builds Chavdar buses at Botevgrad and has negotiated a co-operation agreement with Greece.

The Tatra works at Prague-Smichov is a major builder of trams. The old Franz Ringhoffer plant was an important producer before the war but the firm rose to greater prominence when it acquired a licence in 1948 for the manufacture of cars to an American design. Exports increased immediately but progress has been maintained with new designs. The new three-link tram incorporating hinges and stretch covers is 30 metres long and can accommodate 350 people, resembling a Metro with its interior design and maximum speed of 65 km/hour. Tatra have supplied trams to Eastern Europe since 1952 but Comecon integration has reinforced the export role. Poland produces tramcars at the Konstal works in Chorzow but the GDR's tram factory at Gotha has switched to the production of domestic appliances as the main tramway systems in the country (Dresden, Halle, Leipzig, and Magdeburg) standardize on the Czech tram. Romania has also scaled down her own production in favour of imports from Czechoslovakia.

Railway equipment, aircraft, and shipping

Production of railway equipment has been affected by Comecon integration schemes and this is particularly noticeable for locomotives and some highly specialized coaches and wagons. Each country retains capacities for the more basic types of rolling stock. Czechoslovakia has a strong position in locomotive building and has invested quite heavily in research and development with a railway research institute in Prague which has the use of a test track at Cerhenice (in the Labe valley 50 km from Prague). There are circuits of 13 km and 4 km which can be used to check fuel consumption and also for noise and braking trials. Electric locomotives manufactured by Skoda of Plzen have been supplied to Bulgaria, Poland, and the Soviet Union as well as the home market. The enterprise has developed dual voltage locomotives and a large locomotive for handling heavy trains at high speeds (650 tonnes at 200 km/hour) in the Soviet Union. The latter is a two-section locomotive with a rating of 8,400 Kw which can operate at temperatures ranging from -50 degrees to +40. It will therefore be useful on the new Baykal-Amur line and other railways in Siberia. Meanwhile the GDR rail vehicle industry (Vereinigter Schienenfahrzeugbau) has ceased construction of large locomotives and concentrates on diesel locomotives in the intermediate power range using the Hans Beimler works at Babelsberg near Berlin. (16) Electric locomotives for heavy freight trains (ratings up to 5,400 Kw) are also produced.

    The GDR had a particular reputation for rolling stock through the Waggonbau factories at Bautzen (sleeping cars), Dessau (refrigerated wagons), and Halle-Ammendorf (standard passenger coaches). Both the GDR and Poland have made use of the Cerhenice test track in respect of new types of rolling stock. Romania meanwhile has retained its own capacities for both diesel and electric main-line locomotives. Western designs (Asea and Brown-Boveri) have been used to satisfy a range of requirements. There is no co-operation within Comecon in this respect: only an arrangement between Electroputere of Craiova (where a large electrical engineering complex developed out of steam locomotive repair shops after the Second World War) and Rade Koncar of Zagreb regarding the production of six-wheeled and four-wheeled bogie electric locomotives respectively. Romania's 23 August engineering works has developed a light diesel locomotive for agro-industrial use. Its low fuel consumption, less than that of a lorry with the same power (100 hp), and very low weight of 14 tonnes, suitable for track without a proper foundation, opens the way for efficient use of internal railways by large enterprises in food processing and light industry. Hungary has also shown enterprise in developing the Ganz-Mavag railbus, a reliable vehicle using a simple technology which has won export orders in the Third World. The same firm has also exported trains to New Zealand. Bulgaria builds coaches at Dryamovo and has facilities for wagons at Burgas, Cherveno, Sofia, and Zname (also Karlovo and Ruse for wagon repairs).

There was a significant aircraft industry in Eastern Europe during the inter-war years (especially in Germany) but production did not always continue after 1945 given the reduced priority for air transport. Factories were transferred to other uses concerned with engineering and vehicle building. There are however cases of both continuity and revival, involving both collaboration within Comecon and links with western firms. Aircraft production in Czechoslovakia dates back to 1918 when Rudy Letov opened his factory in Prague. After several relocations the business settled down in Prague-Letnany, where the main work is concerned with automatic control equipment and flight simulators for the Prague-based Aero trust whose main installation is at Vodochody where MIG aircraft are assembled under Soviet licence. Recently the industry has gained widespread recognition for its small aircraft. These include training aircraft, used widely by the Warsaw Pact air forces and a number of Third World countries, executive-type aircraft for small numbers of passengers, and aircraft for such special purposes as aerobatics and crop-spraying.

Production is co-ordinated between factories at Vodochody and Kunowice (the latter in Moravia) and engines are made at the Motorlet works in Jinovice with a tradition in engineering dating back to 1911. Capacity to operate in extreme climatic conditions means that there is some potential for export to the Soviet Union. By contrast Romania has favoured western collaboration following the decision of 1968 to rebuild the small infant aircraft industry important during the 1930s but abandoned after the war. A co-operation agreement with British Aerospace led to the purchase of 1-11 aircraft and, in 1979, a contract covering assembly of the aircraft in Romania using British equipment initially but later on an independent basis. This ROMBAC 1-11 is now being used for passenger and freight transport in Romania. Deals have also been made for a Romanian contribution to the Britten-Norman Islander aircraft, which will provide experience in the production of commuter transport aircraft and various military versions, and with French companies for the production of Alouette and Puma helicopters. There is also joint production of military aircraft with Yugoslavia.

The ship-building industry has been a dramatic indicator of Poland's strong maritime position arising out of the territorial changes of 1945. The industry has been hailed as a stimulus for technical and organizational progress and hence the more rapid economic development of the country; (17) but heavy investment is required to ensure that the yards remain competitive. Poland has therefore modernized its shipyards at Gdansk-Gydnia and Szczecin and various factories producing equipment. In 1974 the Comuna Paryska yard in Gdynia launched a 105,000 tonne oil-and-ore-carrying vessel ordered by the USSR and ships of 200,000-400,000 tonnes were then envisaged for the future. However Poland's success in developing a ship-building industry has given rise to some dilemmas in the 1980s arising out of the need to modernize the domestic fleet. Should capacity in the shipyards be allocated to meet the needs of Poland's own merchant shipping requirements or should foreign orders be given priority, even at the expense of having to go abroad to get the country's own needs serviced rather than contemplate the increasing obsolescence and inefficiency of the fleet? Since export orders are frequently on credit and the Poles can earn hard currency by operating new ships themselves, there is a strong case for compromise.

Russian orders have also been crucial in maintaining the ship-building industry on the Danube at Komarno in Slovakia where Amur-type ships are built suitable for bulk trade (coal, grain, or timber) and containers on Soviet inland waterways. Bulgaria too has benefited from Soviet support. The ship-building industry arose out of the Bulgarian-Soviet joint company Korbso set up in Varna in 1950 for building and repairing (a drydock opened in 1955). The Soviet interest was withdrawn in 1954 and the yard continued to function as a purely Bulgarian ship-building enterprise, reconstructed in 1962 to cope with oil tankers of 100,000 tonnes and ore/bulk carriers up to 38,000 tonnes. The Ivan Dimitrov shipyard at Ruse (opened in 1959) specializes in 1,700 hp vessels for pushing barges on inland waterways and 1,500 tonne liquid fuel bunker vessels for use at major ports. Ruse also builds oil tankers which can be used on both seaways and navigable rivers: 2,150 tonne vessels of 1,400 hp and 5,000 tonne vessels with 2,000 hp. Construction of marine engines involves production facilities in a number of Bulgarian cities including Novipazar, Provadia, Shumen, Tirgovishte, Tolbukhin, and Varna. Many of the orders are placed by the Soviet Union. Yugoslavia meanwhile built up an important ship-building industry (mainly at Rijeka) geared to the needs of her own fleet as well as western and Third World orders. However the industry has been badly hit by the depression and the growth of tonnage in the domestic fleet has been very small. Faster expansion is planned for the future with a growth of containerized services, which currently operate (mainly from Rijeka) to the Far East, Middle East, and North America, and greater promotion of Yugoslav ports in other Balkan and Danubian countries. Romania has followed a roughly similar strategy in development of her ship-building industry which covers river boats at Orsova and Drobeta-Turnu Severin and sea-going ships at Constanta, Galati, and Mangalia.

ENERGY: THE PROBLEM OF LIMITING TRADE DEPENDENCE

Energy resources are unevenly distributed both between and within countries despite post-war prospecting, which has led to important discoveries in formerly deficient regions. Eastern Europe is relatively well off for coal, although most of the reserves consist of the lower-grade coal, especially

Figure 1.3 Power stations and electricity transmission in Eastern Europe c. 1940

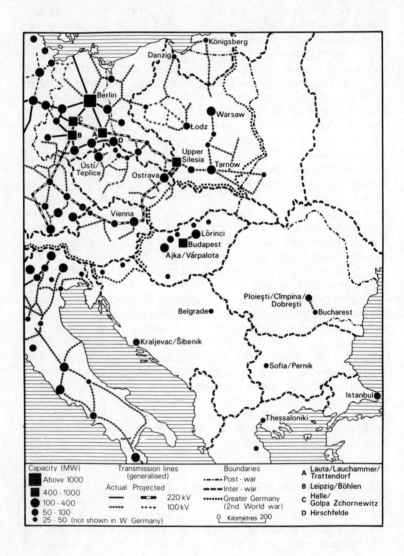

Source: National atlases.

lignite. Poland is particularly well endowed with hard coal however, and until 1979 she was the world's second largest coal exporter (after the USA). Labour difficulties (arising out of pressure on miners to work a six-day week) meant a loss of output, but coal remains Poland's main export commodity. Hydrocarbons are not nearly so plentiful however. In the early post-war years Romania had a surplus of both oil and gas; at the present time Albania's new posture of 'self-reliance' is greatly facilitated by exports of oil and oil products, while the development of hydro-electricity has outstripped domestic requirements and allowed exports to Austria, Greece, and Yugoslavia (see Figure 1.3).

The search for hydrocarbons has been maintained diligently throughout Eastern Europe but with only modest success. Bulgaria first produced oil in 1954 near the village of Shabla (Tolbukhin) and more extensive prospecting was rewarded by the discovery of oil in 1962 at Dolni Dabnik near Pleven, where a processing plant was built. However the deposits proved smaller than expected and the projected annual output of 1 million tonnes has not been achieved (0.5 million tonnes were produced in 1967 but output during the decade fluctuated around the 0.4 million tonne level). Hence the need for imported crude and the decision of 1958 to build a refinery at Burgas (where capacity has risen from 1 million to 12 million tonnes) followed by a refined products' pipeline to Sofia. The search for natural gas has also been disappointing, for the only major deposit was found in 1963 at Chiren near Vratsa where a production unit for urea was set up. Once again the disappointment led to arrangements in 1969 for the import of Soviet gas by pipeline beginning in 1974. Elsewhere there have been modest finds. Thus Hungary's Zala oilfield has been supplemented by new discoveries at Nagylengyel in the 1950s (so production reached 1.8 million tonnes in 1963) and further prospecting has been done at Budafapuszta and Lovaszi. Natural gas comes from Zala and from new finds on the Great Plain at Hajduszoboszlo. At present there is offshore prospecting for oil in both the Adriatic and the Black Sea (Romania had four rigs in service in 1985 with two others on order) and prospecting on land has reached depths of 7,000 metres in Czechoslovakia.

The import of hydrocarbons has therefore been unavoidable with most of the supplies coming from the Soviet Union (Table 1.3). For a time there was a rapid

Table 1.3 Energy, production, and trade 1960-80

Country	Total primary energy production (in million tonnes coal equivalent)						Coal production (million tonnes)					
	1960		1970		1980		1960			1980		
	A	B	A	B	A	B	AA	AB	B	AA	AB	B
Bulgaria	7.7	+1.8	16.1	+15.5	17.1	+50.2	0.6	15.4	*	0.3	29.8	+6.0
Czechoslovakia	49.8	+4.6	63.8	+15.5	66.4	+36.6	26.4	57.9	+0.7	28.2	93.1	-0.4
GDR	73.6	+10.4	81.0	+15.0	83.6	+42.9	2.7	225.5	+8.1	0.0	258.0	+14.9
Hungary	14.7	+3.9	20.7	+10.2	22.0	+21.0	2.8	23.7	+1.8	3.1	23.4	+4.3
Poland	93.4	-8.3	130.6	-12.6	175.0	+5.4	104.4	9.3	-22.2	193.1	36.9	-27.6
Romania	34.5	-8.8	64.8	-1.4	85.7	+19.4	3.4	3.4	+0.4	8.1	27.1	+6.8
Yugoslavia	15.5	+0.3	21.4	+8.7	28.6	+21.8	1.3	21.4	+1.2	0.4	46.6	+4.5

	Oil production (million tonnes)						Gas production (million cubic metres)					
	1960		1970		1980		1960		1970		1980	
	A	B	A	B	A	B	A	B	A	B	A	B
Bulgaria	0.2		0.3	+0.9	0.3	+13.3			473		na	na
Czechoslovakia	0.1	+2.3	0.2	+9.8	0.1	+19.3	1,443	-5	1,204	+1,271	na	na
GDR	0.2	+1.8	0.3	+10.3		+21.8			1,230		na	na
Hungary	1.2	+1.4	1.9	+4.4	2.0	+8.3	342	+186	2,469	+200	6,127	+3,283
Poland	0.2	+0.7	0.4	+7.0	0.3	+16.3	541	+241	4,975	+1,002	5,958	+5,241
Romania	11.5		13.4	+2.3	11.5	+16.0	10,330	-204	25,050	-200	35,171	+1,258
Yugoslavia	0.9	+0.5	2.9	+4.5	4.2	+10.0	52		977		na	na

Source: Statistical yearbooks.

Notes: A Production; AA Hard coal; AB Soft coal; B Trade (- indicates export).

increase in dependence on imported energy. In Czechoslovakia the import of hydrocarbons in the 1960s led to a coal surplus which allowed for the closure of the least profitable coal mines: more than a hundred operations were slated for closure between 1964 and 1970 and some 20,000 jobs were lost.

But the price of Soviet oil supplied to Eastern Europe rose dramatically from 28 dollars in 1975 to 161 in 1982. This reflects the rise in world prices but also the increase in the proportion of the world price which the Soviet Union charges its allies (35.7 per cent in 1975 but 66.2 in 1982). Of course payment is not normally made in dollars since exchanges are arranged on a barter basis but the increases are still felt in the volume of production required to meet the cost. The deterioration in the terms of trade with the Soviet Union can be seen dramatically in the case of oil imports by Czechoslovakia. While the volume doubled during the 1970s from 9.40 to 18.80 million tonnes the value increased 9.6 times from 1.15 to 11.09 billion crowns. Some benefits however accrue from the fact that Soviet oil is still sold at preferential prices, which means a modest subsidy to the East European economies, while the supply of equipment for the development of new oilfields attracts delivery of quantities of crude oil at even more advantageous prices.

Another problem for Eastern Europe has been Soviet insistence on capital contributions towards the cost of oil and gas developments because, at early 1970s prices, it was necessary to invest three times as much capital in oil and gas field developments as was needed to provide manufacturing capacity in Eastern Europe to pay for fuel deliveries. The Russians also realized that oil and gas sales to the west would yield sophisticated machinery that would modernize Soviet industry. This brings some advantage to Czechoslovakia through the passage of Soviet natural gas in transit to the west. Since the first agreement in 1970 a total of four pipelines have been built across the country with a combined length of some 4,200 km. Total capacity is 68 billion cubic metres (rising to 72 in 1992). Czechoslovakia benefits through involvement in the construction and maintenance of the system including large compressor stations like the one at Breclav. She also receives considerable gas supplies in lieu of transit payments and these can be used in trade: some gas is sold to Austria in return for ammonia.

Increasing domestic coal production

Eastern Europe has been forced to cut its imports of hydrocarbons, especially oil. Hungary for example managed to reduce oil imports from 9.6 million tonnes in 1979 to 8.3 in 1980. The consequence of this, given the delays over the nuclear programme, has been a renewed emphasis on coal, despite some labour shortages and the ecological problems which arise out of the strip mining and the emission of dust from power stations and heating plants. In Czechoslovakia it is interesting to see how the present leadership finds it convenient to blame the switch to oil in the 1960s on the architects of the 'Prague Spring' of 1968, although the policy was a logical one at the time and one which has been revised only because of unforeseen price rises, coupled with the Soviet interest in diverting supplies to the west. New investment is now going into the quarries of north Bohemia where measures are being taken to prevent landslides (as quarries push further into hillsides) and to safeguard the mineral waters, especially those of Karlovy Vary which are threatened by the new quarry of Cankov. The old town of Most is being cleared to vacate a site for lignite working and a new town for 60,000 people will have to be built elsewhere. A new generation of walking dragline excavators is being built: the Klement Gottwald metallurgical and engineering works of Vitkovice has a planning and design establishment for mining machinery at Teplice and giant machines weighing 4.5 thousand tonnes are assembled at Bilina. They then walk to the quarries at the rate of about 4 kilometres per month.

In the GDR the effective freezing of Soviet oil exports to Eastern Europe in 1980 brought an immediate response in terms of domestic fuel output. GDR coal output was 258 million tonnes in 1980 compared with only 250 planned. The tenth SED congress called for the conversion of the entire economy to consumption of energy based on low-grade coal. (19) Production is to reach 295 million tonnes in 1985 and 300 in 1990. This will involve severe problems as more land is worked over and the volume of over-burden increases with quarrying at deeper levels. Between 1980 and 1985 the weight of over-burden per unit of coal extracted increased from 3.9 to 4.5 (as much as 7.1 in some pits). Deeper mines also give rise to greater drainage problems and fuel with a higher moisture content. The main complex is Senftenberg, producing 185 million tonnes in 1984. The largest single

59

mine in the GDR is in this area (Nochten) and new workings are to be started at Reichwalde and Scheibe. The village of Erythra near Leipzig has recently been moved to make way for lignite quarrying but this is just one of some fifty settlements which have been disturbed in this way since 1928. Many railways and roads have been realigned. Meanwhile Poland has cut back her planned oil consumption (previously set at 35 million tonnes for 1990) and expects to follow the sequential pattern of brown coal development from Konin to Turoszow and Belchatow with a new development at Legnica. (20) Unfortunately the great depth of the seams and the existence of water-bearing strata have delayed the opening of the Lublin coalfield.

Romania has made a powerful drive to boost coal output from the strip mines in Oltenia. Six mines have been opened up in the Motru field, at Albesti, Horesti, Luerda, Lupoaia, Plostina, and Rosiuta and four units at Rovinari: Beterega, Cicani, Girla, and Tismana. Branch railways have been built from Tirgu Jiu to Rovinari, and from Strehaia to Motru, where a new town is being built. At Rovinari diversion of the river Jiu, to allow mining under the former river bed, has enabled a large artificial reservoir to be created as a fish pond and a major source of industrial water. The lignite will be consumed mainly by power stations at Rogojelu, Turceni and Craiova. Output has risen rapidly although it has been difficult to meet ambitious plan targets. Equipment has been slow in arriving and labour shortages prompted nation-wide enquiries (known as 'Action 7,000') to recruit 7,000 extra workers for the 1980 and 1981 seasons. There have nevertheless been labour difficulties at Motru (especially in 1981), associated with bad working conditions, low pay, and poor food supply. This trouble came after previous unrest in the Jiu valley bituminous coal-mines in 1977. Severe winter weather can also disrupt strip-mining operations through waterlogging and freezing. Evidently the local management has been under the greatest pressure from the ministry to maximize output. Meanwhile Bulgaria's main effort is being made at the Maritsa-Iztok (Maritsa East) lignite mining and power complex with the prospect of new working on the frontier with Yugoslavia at Stanyano-Mazgosh. Coking coal has been found near the resort of Balchik.

Energy links with the Soviet Union (21)

Despite best efforts Eastern Europe will continue to require large imports of fuel and electricity. Imports of hydrocarbons from the Soviet Union arrive by pipeline which is cheaper and more reliable than rail transport. Oil is transported through the Druzhba (Friendship) pipeline completed in 1964. It runs from the Tatar ASSR to Mozyr in Belorussia and then bifurcates with a northern arm running to the Plock (Poland) and Schwedt (GDR) refineries and on to the chemical complexes in the Halle-Leipzig area; and a southern branch running through Czechoslovakia to reach the Bratislava and Szazhalombatta (Hungary) refineries and chemical complexes in North Bohemia. Oil is also sent from Odessa and Novorossiisk to Burgas in Bulgaria and Omisalj (Krk) in Yugoslavia, although the latter country is also supplied by the Druzhba pipeline and the connecting Adria pipeline when there is adequate capacity available. Yugoslavia also imports oil from Iraq via Ceyhan terminal in Turkey. The gas pipeline (Fraternity) was laid in the early 1970s following the negotiation of contracts (in 1968) for the supply of Soviet gas to Austria, FRG, and Italy in addition to Eastern Europe and (in 1970) for transport of gas in transit through Czechoslovakia; 1,200 mm pipes were laid from the Ukraine (Dashava and Shebelinka) to the Czechoslovak frontier at Uzhgorod and from there to Plavecky Peter, north of Bratislava, where the pipeline bifurcates. The northern arm ran north-westwards to Zlonice, north of Prague, where there was a further divide to supply both the FRG via Rozvadov/Waidhaus and the GDR at Hora Svate Kateriny (beginning in 1973). The southern branch supplies Austria (1972), Italy (1974), and France (1976).

However the gas sources have changed over the years as new fields have been opened near Orenburg and subsequently in Siberia at Tyumen and most recently at Urengoy/Yamburg. Capacity has been increased by construction of additional pipelines: 1978 saw the opening of the Soyuz (Union) pipeline bringing the Orenburg (later Tyumen) gas across Czechoslovakia to Rozvadov and Zlonice, using the same route as the first pipeline (apart from a short cut taken from Breclav to the FRG frontier). This further supply arose out of the Comecon programme of economic integration which was agreed in outline in 1971 and worked out in detail by 1974. The diameter of the pipes

61

varies from 1,420 mm in the USSR to 800 mm in Rozvadov. Finally in 1984-5 a third (Transit) pipeline of 1,400 - 1,000mm was laid across Czechoslovakia bringing gas from north-western Siberia. There have been complex negotiations involving East European participation in development work in Siberia and the transmission of Iranian gas to the USSR. The flow of 54.0 billion cubic metres per annum (1984) is adequate for normal requirements but storages are made (for example at Lah in Slovakia where 1.1 billion cubic metres are held underground) to meet winter demand. Other pipelines were built during the 1960s and 1970s to take Soviet gas to Warsaw (520mm pipeline laid in 1966), Sofia (via a 1,020mm pipeline laid across Romania in 1973) and Leninvaros (by means of an 820mm pipeline). The Bulgarian distribution ring has developed out of the local Chiren-Vratsa pipeline of 1966 and connects Pleven, Sofia, Stara Zagora, Karnobat (with a branch to Burgas) and Devnya. Thus it supplies the main installations in the chemical and metallurgical industry (Dimitrovgrad, Kremikovtsi, Pernik, Stara Zagora, Varna-Devnya, and Vratsa) and important units producing cement and ceramics. Also Romania has now plugged into the transit pipeline and receives 1,500 million cubic metres annually.

The first joint venture in construction of an electricity grid involved Czechoslovakia and Hungary in 1953. In 1962 the Mir (Peace) grid came into operation connecting the four northern countries of Eastern Europe with the western regions of the USSR. Romania was connected in 1963 and Bulgaria in 1967, when a unified power system and control centre came into existence. Transfers of electricity were restricted as long as voltages were limited to 220-240 kV but capacity has been stepped up. Bulgaria's first link with the Romanian grid at Craiova in 1967 had a rating of only 220kV but five years later a 400kV line was built across Romanian territory to connect Bulgaria with the Soviet Union: this has contributed to the concentration of industry in the north-eastern part of the country. A link has been made between Poland and the GDR across the Oder. At present 750kV lines are being built to handle large transfers of power from power stations (mainly nuclear) in western regions of the USSR. The first such line was completed in 1978 between Vinnitsa and Albertirsa (Hungary) extending a line already in existence between the Donbas and Vinnitsa. In 1980 power from the Chernobyl nuclear power station was exported to Hungary by this route. Bulgaria also has a link with Turkey opened in 1975 (though planned in 1965) and there is a considerable exchange of electricity between Bulgaria and Yugoslavia. A 110kV link between Breznik and Vurla became operational in 1961 but further links have been installed between Koula and Bor and between Petrich and Strumitsa and there is a 220kV link between Stolnik and Nis. Yugoslavia is connected to her other neighbours: power is transmitted through Yugoslavia from Italy to Austria and Romania and to Italy from Albania. Hungary also has a close link with Austria: surplus from Austrian hydro-power stations is transmitted to Hungary while power from Hungarian thermal stations goes to Austria, when low water conditions reduce the output from the hydro plant.

NOTES

1. Brown and Neuberger (1968); Vajda and Simai (1971).
2. Dewar (1951); Marer (1972).
3. G.W. Hoffman (1985) 'Eastern Europe: fifty years of changes and constraints', Wilson Center East European Program Occasional Paper 1.
4. Kaser (1967); Kohler (1965); Mellor (1971); Patyk (1966); Schaefer (1972). See also A. Broner (1975) Economic Integration in Eastern Europe, Princeton University, NJ: Ph.D. thesis; D.A. Kobal (1974) Comecon and the Warsaw Treaty Organisation: Their Political Role since 1953, American University, Washington, DC: Ph.D. thesis.
5. P. Wiles and A. Smith (1978) 'The convergence of the CMEA and the EEC', in Shlaim and Yannopoulos (eds) 71-106 (quotation p.71).
6. Ibid. p. 76. See also J.M. Montias (1969) 'Obstacles to the economic integration of Eastern Europe', Studies in Comparative Communism 2(3-4): 38-60.
7. Ausch (1972); Kiss (1971); Wilczynski (1974). See also Z.M. Fallenbuchl (1973) 'Comecon integration', Problems of Communism 22(2): 25-39; M.C. Kaser (1972) 'Comecon and the new multilateralism', World Today 28:149-62.
8. H. Trend (1973) Economic Cooperation among Czechoslovakia, East Germany and Poland in the 1970s, Munich: Radio Free Europe.
9. M. Florescu (1973) Twenty years of Progress of Socialist Romania's Chemical Industry, Bucharest:

Meridiane.

10.   Holzman (1987). See also M. Baumer and H.-D. Jacobsen (1981) 'CMEAs Weltpolitik between global limitations and all European potentials: US Congress Joint Economic Committee', East European Economic Assessment: Regional Assessment, Washington, DC: US Government Printing Office, 872-86; K. Botos (1982) 'On the further development of the currency and financial system of the CMEA', Soviet Studies 34: 228-53; L. Csaba (1988) 'CMEA in 2000', Nordic Journal of Soviet and East European studies 5: 11-31; L. Csaba (1985) 'Joint investments and mutual advantages in the CMEA: retrospection and prognosis', Soviet Studies 37:227-47; V.V. Kusin (1982) 'CMEA's economic strategy for the eighties', Radio Free Europe Background Report 127; P. Marer (1977) 'Prospects for integration in Eastern Europe: the CMEA', in Triska and Cocks (eds) 217-37.

11.   A. Caranfil (1971) 'Patterns in the evolution of urban transportation in Romania', Revue roumaine: geographie 15:117-23.

12.   J.F. Tismer (1985) 'The strategy of intensification of freight traffic in the GDR', in Ambler et al. (eds) 189-205.

13.   M. Madeyski and E. Lissowska (1975) 'Problems of the integrated organization of transport', Geographia Polonica 32: 43-52; M. Potrykowski (1985) 'Some problems of the pattern of the transport networks of Poland', Geographia Polonica 51:275-83.

14.   Z. Taylor (1984) 'The diffusion of railway network in Poland as a space-time process', Geographia Polonica 50:75-87.

15.   F. Mohrmann (1977) 'Development and problems of road transport between the East and the West', GeoJournal 1: 65-90.

16.   E.J. Cornwell (1967) 'East German rolling stock industry's export boom', Modern Railways 23: 373-5; 430-4.

17.   J. Zaleski (1978) 'Poland's maritime economy', GeoJournal 2: 179-90 (see p. 187).

18.   Hoffman and Dienes (1985). See also L. Dienes (1976) 'Energy prospects for Eastern Europe', Energy Policy 4: 119-29; J. Gurney (1978) 'Energy needs in the Balkans: a source of conflict or co-operation?', World Today 34(2): 44-51; G.W. Hoffman (1983) 'Energy dependence and policy options in Eastern Europe', in R.G. Jensen et al. (eds) Soviet Natural Resources in the World Economy, Chicago: Chicago

University Press, 659-67; C.J. Joyner (1976) 'The energy situation in Eastern Europe: problems and prospects', East European Quarterly 10: 495-516; J.M. Kramer (1985) 'Soviet-CMEA energy ties', Problems of Communism 34(4): 32-47; S. Wasowski (1969) 'The fuel situation in Eastern Europe', Soviet Studies 21:35-51; B.J. Zajac (1978) Energy Problems and Programs of the CMEA, University of Pennsylvania: Ph.D. thesis.

19. C.B. Schultz (1983) 'The energy dilemma confronting the GDR', Radio Free Europe Background Report 48.

20. J. Kozacki (1978) 'Changes in the geographical environment as a result of open mining', Geographia Polonica 41:81-7.

21. M.B. Green and M.J. Sagars (1985) 'Changes in Soviet natural gas flows 1970-1985', Professional Geographer 37:310-9; D. Park (1979) Oil and Gas in Comecon Countries, London: Kogan Page.

Chapter Two

NON-COMECON TRADE AND CO-OPERATION

It has been seen in the companion volume how Eastern Europe occupies an intermediate position between advanced economies, requiring the stimulation of exports, and developing economies, which need imports of technology and capital goods. The socialist economies need to maximize output from their own industries which have achieved a considerable degree of sophistication but in the 1950s foreign trade was merely an accessory to essentially autarkic development plans. There were undoubted successes but self-sufficiency could not eliminate the technology gap between east and west completely and it could not generate significant economies of scale. Moreover an ambivalent attitude to foreign trade, acceptable only in the last resort to assure the required inputs or eliminate surpluses, resulted in some quite remarkable anomalies: outlays made on the purchase of raw materials and fuels for metallurgical industries (notably in the GDR and Hungary) have been greater than the cost of buying the finished metal while some manufactured goods have been exported for lower prices than those paid for the imported raw materials. (1) Foreign trade has now assumed a much more positive role and it is accepted that 'international trade can become an engine for growth' (2) since export markets can stimulate technical progress and provide economies of scale in those industries selected for specialization. Exports are no longer regarded as ad hoc sales of surplus output. Production is planned with international trade in mind and state trading organizations provide the link between the foreign customers and local enterprises. Thus the Polish Hortex organization (set up in 1958) handles exports of fruit and vegetables (in frozen or various processed forms) to the USSR and western markets (chiefly the FRG). Of course there is a difficult transition period while a specialization strategy is worked out, requiring reallocation of premises and disposal of surplus equipment, and prices are modified to reflect world market values more closely. But there is no doubt that trade has increased sharply and it would be difficult to deny that the recent growth of foreign trade in Eastern Europe has been above what would have resulted from a simple extrapolation of previous trends (see Table 2.1). It can only be ascribed to a conscious policy change dictated by the necessities of the growth process itself. (3) It is also evident that much of this increase has arisen from developing trade contacts with the advanced countries of western Europe, North America, and Japan and with the Third World. This chapter concentrates on the extra-Comecon dimension. (4)

THE SIGNIFICANCE OF EAST-WEST TRADE

Growth of east-west trade would appear to be in the interest of all parties. There are political advantages to Eastern Europe in reducing dependence on the Soviet Union, though the option (which is anyway conditional on suitable exports being found) has become less attractive through rising raw material and energy costs. But since there are these constraints on the development of trade contacts there is no reason why the Soviet Union should be alarmed: rather she may perceive tangible benefits through the improved Comecon access to western markets to compensate for any political risks. Comecon countries may become members of the International Monetary Fund, which provides credits on favourable terms to compensate for temporary balance of payments deficits: Romania joined both the IMF and the World Bank in 1972 in order to increase its hard currency borrowings. For the west on the other hand trade with communist countries is of more limited importance and dependence is considerable only for a few raw materials, notably chrome, where the proportion rises to one-quarter of the total demand. Firms have not always been keen to get involved with Eastern Europe's monopolistic state trading corporations but trade fairs have multiplied and each Comecon state now stages regular events (in Brno, Bucharest, Budapest, Leipzig, Plovdiv, and Poznan) which make it easier for western businesses to

Table 2.1 Growth direction and structure of foreign trade 1960-80

Country	a Growth of trade (thousand US dollars per capita) Trade in			b Direction of trade (per cent) Trade with USSR (A), other Comecon (B), and the west (C)								
	1960	1970	1980	1960 A	1960 B	1960 C	1970 A	1970 B	1970 C	1980 A	1980 B	1980 C
Bulgaria	0.15	0.45	2.26	53.1	27.4	13.2	53.0	21.9	16.6	50.8	18.7	12.8
Czechoslovakia	0.27	0.52	1.96	34.4	29.4	17.8	32.5	31.8	22.4	35.8	33.4	22.9
GDR	0.26	0.55	na	42.8	24.8	21.1	39.1	28.2	24.4	35.5	29.6	27.4
Hungary	0.19	0.47	1.67	29.5	32.8	20.7	33.6	28.6	24.5	28.5	23.5	34.7
Poland	0.10	0.22	1.01	30.3	27.0	29.1	36.5	26.8	27.1	32.2	22.6	34.1
Romania	0.07	0.19	1.15	40.1	26.5	21.6	27.0	21.9	34.8	17.0	20.4	38.2
Yugoslavia	0.07	0.22	1.08	7.9	20.7	58.0	9.6	16.9	62.0	21.6	14.3	46.0

Country — c Structure of trade (per cent)

Shares for food (A), raw materials and fuels (B), and manufactures (C) in

Country	1960						1970						1980					
	Exports			Imports			Exports			Imports			Exports			Imports		
	A	B	C	A	B	C	A	B	C	A	B	C	A	B	C	A	B	C
Bulgaria	56.4	9.2	34.4	16.7	24.3	59.0	43.4	8.1	48.5	15.9	29.1	55.0	na	na	na	na	na	na
Czechoslovakia	10.4	19.1	70.5	37.1	27.9	35.0	7.3	18.6	74.2	24.1	23.5	52.4	1.8	3.4	94.8	4.7	60.3	35.0
GDR	5.9	15.7	78.4	39.2	38.5	22.3	7.4	10.1	82.5	28.1	27.6	44.3	6.4	14.8	79.8	18.9	36.7	44.4
Hungary	27.4	12.8	59.8	29.2	27.7	43.1	26.7	14.4	58.9	24.4	23.6	52.0	62.3	0.7	37.0	8.1	48.6	43.1
Poland	23.0	34.0	43.0	33.9	25.3	40.8	15.9	23.9	60.2	21.4	26.6	51.8	3.5	13.8	82.7	9.9	54.7	35.4
Romania	35.9	36.9	27.2	18.4	34.3	47.3	26.8	22.7	50.5	15.6	30.4	54.0	12.0	36.2	51.8	9.0	57.0	34.0
Yugoslavia	na	na	na	na	na	na	18.7	10.6	70.7	7.0	15.7	77.3	5.3	1.6	93.1	6.7	61.5	31.8

Sources: Radio Free Europe Research and United Nations yearbooks. Due to slight differences in the method of calculation, information for 1980 (except Romania) and for Yugoslavia in 1970 is not strictly comparable with the rest.

make contact with customers. Czechoslovakia's trade fairs began in 1955 and they have been held annually in Brno since 1959, supplemented by an agricultural exhibition held annually at Nitra since 1973.

The growth of trade has been enthusiastically welcomed by western financial institutions. Financing deficits has been seen as good business because of Eastern Europe's good repayment record, backed up by the belief that there is ultimate security in Russian gold. There is also some reassurance in the presumption that Comecon countries with their central planning system can deal more effectively with economic stress by reducing the supply of goods to the home market and cutting back on imports than can the market economies of the western democracies: thus even during recession the East European economies have continued to grow and this maintains their credit-worthiness. Certain countries gave grounds for concern with above-average levels of debt in relation to hard currency earnings but money continued to be available at attractive interest rates until the Polish crisis of 1981-2.

. East-west trade has certainly helped to reduce Eastern Europe's isolation and the increased contacts provide opportunities for western influence to be exerted. (5) Individual governments can use trade contacts to press for reform in Eastern Europe. The USA has certainly used trade as a lever for modest internal reform. In dealings with Romania for example there have been important agreements on copyrights, patents, and marine transport, and the 'General System of Preferences' was granted in 1976. But the most important issue has been the granting of Most-Favoured-Nation (MFN) treatment, first in 1975 but subsequently on an annual basis. This has discouraged long-term trade agreements but has brought some flexibility on issues concerned with emigration and other human rights. In 1983 negotiations over renewal of MFN status resulted in concessions over the tax which the Romanians had previously imposed on prospective emigrants to cover the cost of the education they had received.

The case of Japan

Western involvement has increased and the Japanese case is typical. During the late 1950s and 1960s Japanese activity was concerned largely with restoring diplomatic relations

70

and increasing the turnover of trade. Diplomatic relations were established in 1952 (Yugoslavia), 1957 (Czechoslovakia and Poland), 1959 (Bulgaria, Hungary, and Romania) and 1973 (GDR) while the first trade agreements were signed in 1958 (Poland), 1959 (Czechoslovakia and Yugoslavia), 1960 (Romania), 1961 (Bulgaria and Hungary), and 1975 (GDR). Since 1970 economic relations have been extended by the creation of bilateral business and financial institutions, the establishment of joint companies and negotiation of substantial Japanese credits which have helped to bring about a great increase in trade. Joint (non-governmental) economic committees were set up with all countries in 1972 while mixed governmental committees emerged first with Poland (1967) and rather later with Romania (1975), Hungary (1977), and with Bulgaria and Yugoslavia (1978). The first joint companies were reported in 1968 (Bulgaria), 1971 (Poland), 1978 (Czechoslovakia and Romania), and 1980 (Yugoslavia).

THE IMPACT OF WESTERN TRADE IN EASTERN EUROPE

The East Europeans have adopted very different stances. Romania has maintained a very high priority on industry equipped with western plant. This strategy has involved heavy sacrifices over consumption but on the other hand some Western economists argue that 'Romania's use of capitalist markets for her hard-goods exports at an early stage of development is likely to provide a basis for more successful exporting of manufactured goods to those same markets at a later stage of development'. (6) In the long run therefore the tendency to over-industrialize, through restricting investment in agriculture and the training of skilled labour and managers, may turn out to be advantageous. However at the present time Hungary has probably gained most from east-west trade because economic reform through the New Economic Mechanism (NEM) has been conducive to maximization of export opportunities. Agricultural exports are prominent including small volume/high value consignments such as game meat (fallow deer, roe deer, and wild boar) which totals some 4.5 thousand tonnes each year; manufactures show a similar diversity with the success of toy exports boosted by the Rubik Cube (the invention of the Hungarian Erno Rubik). Bulgaria has introduced its own mechanism for greater enterprise autonomy, which is appropriate in the context of the country's drive into high technology industry including electronics, robotics, and machine tools. With modest debts outstanding, Bulgaria has become a good prospect for western suppliers especially those which can help the Bulgarians market the products of their new industries. By contrast, Czechoslovakia has followed an extremely cautious policy of borrowing from the west and in the early 1980s this meant relatively modest outlays in servicing debts. (7) But industry may be losing out as regards its long-term export potential. Cars and textile machines which once enjoyed a good reputation have been eclipsed technologically and hard currency earnings are now relatively small.

As for the non-Comecon states Albania follows a maverick course, declining to trade with either of the superpowers while expanding business with Eastern Europe (especially Czechoslovakia) and the west. Her need for technical assistance is even leading to interest in establishing diplomatic relations with the FRG previously ruled out by Tirana's insistence on unrealistic Second World War reparation demands. The need for spare parts and other equipment for installations initially built by the Chinese had led to resumption of trade with her erstwhile benefactor in 1983. Yugoslavia's trade contacts are much more diverse. An agreement was reached with the EEC in 1980 allowing most of Yugoslavia's manufactures to have duty-free access: about thirty sensitive products are free of duty until quotas are exceeded. Unfortunately she is seeking to sell to the Community types of products which are already in good supply. Although concessions have been made over some agricultural products pressures from farming interests within the EEC have imposed limits. Thus the export of 'baby beef' to Greece has been constrained by duties which give French and Italian farmers a crucial advantage. A change to more profitable enterprises is made difficult by Yugoslavia's system of small private farms.

A special facet of east-west trade are the 'Inner-German' exchanges between the GDR and FRG which have been fairly buoyant thanks to the use of accounting units equal to D-marks (which mean that the GDR does not have to spend scarce hard currency). Moreover the GDR's special relationship with the FRG has resulted in the provision of capital to improve infrastructure, especially that concerned with movement between the two German states and between the FRG and West Berlin. The two states have a mutual interest in closer co-operation although they are tied to their respective alliance systems and have limited room for manoeuvre: the GDR wishes to develop trade links while the FRG is keen to increase the scope for travel by individual GDR citizens and is prepared to make economic concessions to facilitate movement and migration. Several important trade deals have been announced at the Leipzig Fair, such as the contract awarded to Hoechst for the construction of a chemical plant at Schkopau in the GDR. The Leipzig Fair has great historical significance (having been in existence since 1165) but it has become a focal point for east-west trade; for businessmen from the FRG it is a gateway to potential markets in Eastern Europe generally.

All round the impact has been considerable. The East European share of all trade with the west has not returned to the inter-war level of some 70 per cent and is unlikely to do so in future. But dependence on the west for investment goods is high and will probably remain so, owing to the lack of alternative suppliers and the technology gap at home which limits import substitution. Technology transfer accelerated during the 1960s and 1970s, initially through the sale of machinery and licences. Throughout the 1960s, with the exception of only one year, there were trade deficits between Comecon and EEC countries, particularly the FRG and France, and the situation remained difficult in the 1970s despite progress with invisible earnings of hard currency (through tourism and transport) because there is little scope for the export of fuels and raw materials. (8) Yet Eastern Europe wants to avoid increasing the proportion of exports earmarked for servicing debts. At the end of the decade these values were very high in Poland (140 per cent), substantial in Romania (50) and the GDR (45) but progressively more moderate in Hungary (35), Bulgaria (30), and Czechoslovakia (20). Consequently much now depends on the ability to assimilate new technology and produce goods which can then penetrate western markets. Performance to date has been disappointing but partly because some western credits negotiated during the 1970s were wasted to some extent by mismanagement and by investment in economically unviable projects linked with over-ambitious growth targets (especially in Poland). (9) Research and development was under-valued in the past but the situation is changing, sometimes stimulated paradoxically by western trade restrictions. Thus American controls on technology exports have encouraged firms in Czechoslovakia (Kovo) and the GDR (Robotron and Mikroelektronik) to expand their own product range.

JOINT VENTURES

Hard currency shortages have given rise to a marked preference for industrial co-operation which provides for payment of western suppliers in goods produced by the eastern partner using the new equipment. (10) Of course the former will expect some reward for risk-bearing (through the forward commitment to purchase at an agreed price) but the arrangement is still attractive to the latter because a bridgehead on the western market is secured. And if the western partner re-sells the goods under his own label or uses the goods as components in an assembly operation, there is some guarantee that the factory in Eastern Europe will be kept technologically up-to-date. It should not be overlooked either that research and development in the Comecon countries is beginning to generate ideas which can be advanced by east-west co-operation, as the western partner may be able to work up certain eastern research results and produce commercially viable products more quickly and effectively than the eastern partner working alone. A modified form of co-operation agreement involves the 'joint venture' which was first provided for by legislation in Romania in 1971 (followed later by Hungary and Poland). This means equity contributions by the two partners with the western partner again transferring technology and benefiting from lower-cost products using relatively cheap East European labour as well as greater control over the use of the equipment or expertise.

Joint ventures have not resulted in really large-scale injections of capital but they have enabled foreign credit to be used efficiently with the support of skilled management and guaranteed markets. Investments have therefore been profitable to both parties. There is always the danger that joint ventures may constrain East European enterprises by locking them into a given technology, which allows little scope for independent innovation. Clearly neither side will want its trade and investment relations to create anything more than a 'limited dependence' on the other. However, most sensitivity has been shown on the political front. Hungary, Poland, and Romania passed legislation which allowed joint economic ventures in the early 1970s but the other three Comecon states maintained strong reservations and only Bulgaria showed signs of shift by the end of the decade. Hungary and Poland (as well as Yugoslavia) have been most active in making co-operation arrangements and on the western side the FRG has been most closely involved: nevertheless the results remain quite modest compared with the level of co-operation within the EEC and developments so far may be seen as a tentative development strategy which attempts to emulate the technology-diffusing characteristics of western multinational co-operations.

Hungary and Poland

Hungarian interest in co-operation is another product of the New Economic Mechanism which allowed certain enterprises to engage directly in foreign trade, rather than operate through the state trading corporations. One agreement between a frozen food firm in the west and a Hungarian state farm has meant the introduction to Hungarian agriculture of quick-freezing and packaging technology while another liaison between two textile firms has allowed the Hungarian partner to obtain production and management expertise (with consequential improvements in productivity) and increased sales on western markets. The American clothing manufacturer Levi Strauss has invested in Hungary, leasing plant from the knitwear factory of Kiskunhalas, while the Belgian Bioprex company is collaborating with the state farm of Csongrad over the breeding of algae (taking advantage of the favourable climatic conditions and the thermal energy available locally).

A significant venture in terms of the modernization and sophistication of East European industry is the joint venture Euroconsult involving the Struktura organization and the Interag foreign trade corporation in Hungary and an engineering consultancy in Switzerland. Struktura is seeking opportunities to market its computer technology and engineering design system up to western standards and so penetrate the market in developing countries where there will be considerable opportunity over the longer term. The aim is to concentrate on aspects of metallurgy where the Swiss partner has particular expertise. Two other unusual agreements date back to 1981. Tatabanya Collieries have made arrangements with a subsidiary of Occidental

Petroleum so that the Hungarian Haldex process can be used to reclaim useful materials from the rock-piles of mines in the USA. Several plants will be set up with joint investment and will be supervised by technicians from Tatabanya. Then there is the deal between a Hungarian foreign trade company Transelektro and various Italian firms to supply power units to Iraq. The Hungarian Ganz electric works will manufacture gas turbines to be assembled in Iraq by Hungarian technicians. (11) A similar arrangement with Italian industry will enable Ganz to build electric generators and auxiliary equipment for Finland.

In Poland the interest in foreign expertise for vehicle construction (underlined by the agreement with Fiat in 1965) is complemented by Austrian and Italian demand for Polish coal and electric power respectively. In Yugoslavia where foreign investment through joint ventures was allowed as far back as 1967 agreements have been made between Sava of Kranj and Semperit of Vienna for the production of tyres, and between Pobeda of Novi Sad and Steyer-Daimler-Puch of Steyer for the manufacture of tractors. It is also worth mentioning that the comprehensive development programme for the Adriatic was drawn up between 1966 and 1969 under United Nations' auspices and involved a Polish-Italian design team which made a particularly important contribution to the transport network, laid out in such a way as to balance the expansion of tourist facilities with the conservation of the historic centres and to make some allowance for industry and the reactivation of depopulated rural areas.

Prospects

Western assistance, and joint ventures in particular, still remain very important for Eastern Europe. The crisis in Poland in the early 1980s was solved in part by the imposition of martial law and disbandment of the Solidarity trade union, and also by help from socialist countries in the form of credits and deliveries of raw materials and manufactures. But Poland's Comecon partners could not supply either spare parts to maintain machinery purchased in the west in the 1970s or equipment needed to complete a large number of industrial projects. Despite the anomalies inherent in the sanctions programme the Polish government certainly feared that the stand taken in the west against the excesses of the martial law regime might leave Poland politically isolated and, in economic terms, back in the position of the 1960s when most of her exports were raw materials and semi-finished goods. The current trade concept involves a growth of exports from Comecon countries coupled with increasingly selective purchases from the west. But postponement of purchases of western products means postponement of reconstruction of industrial plants and increased problems over compatibility in future. (12) Drastic reduction of western indebtedness may be a hollow victory: new projects in Eastern Europe may not be well equipped and this could affect the ability to export over the longer term. Thus it could be argued that after stabilizing their debt it would be desirable for Eastern Europe to maintain a moderate level of borrowing from the west.

Romania also runs the risk of over-reaction to economic crisis. In 1982 there was a serious loss of confidence arising from excessive delays in payment to suppliers, loss of personal contacts as a result of reorganization in the foreign trade ministry following the dismissal of the minister, and complications in negotiations through Romanian insistence on compensation deals, following the law of 1980 which requires any western supplier to accept Romanian goods for all payments (Poland and Bulgaria also take up fairly hard positions on counter-trade while the GDR, Czechoslovakia, and, especially, Hungary are less insistent). Romania has tried to reduce her debt in as short a time as possible and it is interesting to see how balance between exports and imports has changed in dealings with western Europe (see Table 2.2). Imports were 119.4 per cent of exports in the 1960s and 110.4 per cent in 1971-5, but they fell back to 95.1 per cent in 1976-80 and 44.4 in 1981-5. Oil imports have become a crucial issue and, since Romania has tried to avoid very heavy dependence on the Soviet Union, Middle Eastern and African sources have been used. Trade with Asia and Africa accounted for only 11.0 per cent of the total in the 1960s but the share has risen progressively to 17.5 in 1971-5, 25.4 in 1976-80, and 30.6 in 1981-5. Also the excess of imports over exports has increased: 108.3 (1981-5) and 134.2 (1976-80) for Asia; 128.9 (1981-5) and 122.5 (1976-80) for Africa.

Table 2.2 Romania's foreign trade 1960-85

Region and Country (a)	1960-70			1971-75		
	A	B	C	A	B	C
E. Europe (8)	4.33	5.41	94.7	8.52	7.88	92.4
USSR	2.50	2.35	93.9	4.15	3.58	86.2
GDR	0.46	0.46	99.1	1.18	1.20	101.5
Poland	0.27	0.27	100.1	0.78	0.68	87.2
Czechoslovakia	0.58	0.58	100.8	0.99	1.04	104.2
Hungary	0.29	0.20	70.8	0.54	0.53	98.5
W. Europe (17)	1.94	2.32	119.4	6.48	7.15	110.4
FRG	0.55	0.63	114.7	1.77	2.18	122.8
Italy	0.40	0.33	81.0	1.02	0.83	81.3
France	0.20	0.37	182.6	0.64	0.84	131.8
UK	0.19	0.32	166.1	0.60	0.89	148.8
Netherlands	0.06	0.09	142.4	0.43	0.35	80.1
Asia (25)	0.72	0.58	80.5	2.52	2.41	95.6
Iran	0.07	0.05	74.2	0.35	0.45	127.5
Iraq	*	*	*	0.13	0.05	36.1
China	0.26	0.21	81.5	0.77	0.74	96.8
Syria	0.02	0.02	92.9	0.10	0.04	44.4
Japan	0.04	0.09	278.5	0.16	0.49	302.4
Africa (14)	0.19	0.12	63.4	0.79	0.62	79.1
Egypt	0.10	0.09	90.6	0.27	0.18	67.1
Libya	0.04	*	*	0.27	0.15	56.2
Nigeria	*	*	*	0.02	*	*
Morocco	*	*	92.3	0.03	0.09	315.1
Algeria	0.02	0.01	59.7	0.10	0.16	164.9

A	1976-80 B	C	A	1981-85 B	C
16.11	15.96	99.1	60.72	58.34	96.1
7.23	6.97	96.5	32.51	29.42	90.5
2.59	2.66	102.6	7.82	7.98	102.1
1.53	1.55	101.2	6.24	7.00	112.2
1.82	1.81	99.5	4.65	4.41	94.7
1.29	1.28	99.5	4.05	3.52	87.0
11.95	11.36	95.1	53.54	23.79	44.4
3.33	3.22	96.6	13.09	6.32	48.3
1.60	1.10	69.0	11.14	2.33	20.9
1.12	1.44	127.7	5.56	3.14	56.5
0.82	0.95	115.0	4.29	5.10	72.2
1.05	0.64	61.3	3.80	1.25	32.9
6.93	9.30	134.2	34.79	37.51	108.3
0.94	1.75	187.0	5.42	11.91	219.7
0.38	2.46	646.2	9.96	2.88	29.0
1.83	1.86	101.1	6.99	5.45	78.0
0.87	0.36	41.1	1.74	7.21	413.5
0.35	1.00	283.0	1.45	1.81	124.9
2.08	2.55	122.5	10.81	13.93	128.9
0.72	0.32	44.6	5.01	7.74	154.7
0.59	1.10	185.5	2.21	3.96	178.8
0.12	0.39	336.6	0.87	0.88	101.0
0.15	0.17	112.3	0.47	0.42	91.1
0.33	0.24	72.4	1.11	0.19	16.8

	A	B	C	A	B	C
America/ Australia (13)	0.12	0.24	206.1	0.62	1.00	160.9
USA	0.03	0.15	458.3	0.30	0.68	225.3
Cuba	0.03	0.03	76.2	0.06	0.10	166.6
Brazil	0.03	0.02	86.7	0.04	0.07	168.2
Canada	*	*	77.3	0.09	0.03	38.1
Australia	*	*	305.9	0.01	0.03	184.4
Total (77)	7.29	7.35	100.9	18.93	19.06	100.7

Source: Statistical yearbooks.

A   Exports per annum⎫ Figures in billion lei valuta
B   Imports per annum⎭ (1960-80) and billion lei (1981-85)
C   Imports as a percentage of exports
*   Less than 0.01
a   Details are given for the five largest trade exchanges (based on 1981-85 figures) in each area
b   Based on figures for 1960, 1965, and 1970 only

THE THIRD WORLD DIMENSION

The Warsaw Pact states of Eastern Europe, with the exception of Romania, have made a substantial contribution to Moscow's successful penetration of the Third World. This arises from a political concern for solidarity in the post-colonial era, although this consideration would place Yugoslavia's genuine commitment to non-alignment on a different plane from the ambitions of the Soviet bloc countries to seek positive allies. During the Khrushchev era efforts were concentrated on parts of Asia, especially Egypt, India, Indonesia, and Syria. They were able to help orchestrate grievances against former colonial rulers and assist national liberation movements with military support and propaganda offensives against western colonialism and neo-colonialism. Unfortunately many Third World governments rejected any thought of ideological alignment with Moscow and its allies and communists were vigorously persecuted in Egypt and Iraq. In the Brezhnev era help was stepped up (stiffened by Soviet military advisers, Cuban combat troops, and GDR security agents) and targeted on states where pro-Soviet parties had gained power

2.30	3.50	152.0	13.33	10.78	80.9
1.65	2.45	148.7	9.17	6.12	66.6
0.11	0.08	68.1	2.00	1.32	65.7
0.19	0.24	122.8	0.66	1.31	197.3
0.14	0.15	111.5	0.70	0.37	52.1
0.04	0.20	543.6	0.11	0.70	633.1
39.38	42.67	108.4	173.10	144.43	83.4

(Afghanistan, Angola, Congo-Brazzaville, Ethiopia, Mozambique, and South Yemen). There have been increases in East European personnel in Third World countries and while it has been economically most rewarding to extend involvement in oil-rich countries like Iraq, Libya, and Kuwait close attention has also been given to the Caribbean, where the GDR has been particularly prominent in developing relations with Guyana, Nicaragua, and for a time Grenada. In Africa too the erosion of colonialism offered good opportunities in the 1970s. Diplomatic and trade relations have been supplemented by inter-party co-operation and cemented by high-level visits. (13) Trade has not grown very rapidly but the military role has been substantial: Czechoslovakia and the GDR have supplied arms and military personnel. Particular mention may be made of the GDR 'Afrika Korps' comprising some 9,000 military and civilian advisers, which has organized police and security forces in Congo-Brazzaville, Ethiopia, and Mozambique, drilled Palestinian commandos in South Yemen, and supported guerrillas in Zimbabwe (before independence) and South Africa. Altogether the GDR provides two-thirds of all East European aid to the Third World. The other Warsaw

Pact countries therefore contribute at a relatively low level.

Of course there is a wide spread of conventional trading activities across the Third World. Interest in long-term co-operation over minerals and other raw materials, often based on geological surveys, has increased with Soviet reluctance to supply all of Eastern Europe's increasing needs. It offers some prospect of reduced dependence on Comecon in future. In particular relations with the Third World offer the prospect of obtaining scarce raw materials cheaply in exchange for capital and consumer goods for which there is little demand in other markets. They also have political significance in that credits from Eastern Europe strengthen the bargaining position of the developing countries in dealing with the advanced capitalist nations. Much of the trade with the Third World involves prospecting. Czechoslovakia has followed up its successes in locating ferro-nickel reserves in Albania (1956-9) with work in Asian Comecon countries (Mongolia and Vietnam). African countries have also benefited, notably Libya and its oil industry. Reference should also be made to building projects in the Third World like the construction of the Trans-Sahara Highway, involving Algeria, Mali, and Niger. Work by Czechoslovak firms concerns agriculture (irrigation and drainage schemes), power (oil refineries, power stations, and gas processing plant) and transport (roads and bridges) but mainly industry (flour and timber mills and metallurgical/heavy engineering plants). Much of the work has been done by building companies from Slovakia (such as Stavoindustrie of Banska Bystrica and Doprastav and Termsostav of Bratislava) working in Argentina, Brazil, Egypt, India, Iraq, and Syria. Bulgaria's Technoexportstroi Corporation designs and builds a range of installations including airports, hospitals, irrigation systems, sports arenas, and theatres. Activity has been particularly brisk in the oil-rich countries of Iraq (Baghdad airport) and Libya (Tripoli airport and sports stadiums at Benghazi and Tripoli). Hungary has developed considerable expertise in the building and equipping of vocational training schools and workshops. Most business has been done with Africa (Algeria and Nigeria) and South America (Brazil, Ecuador, and Peru).

In relation to her means Romania makes a substantial effort and one which is rather different in scope from that of the other Comecon states. (14) The Romanians share Yugoslavia's genuine commitment to non-alignment which stands in contrast to the Soviet search for positive allies. Romania's influence in the Third World has been increased by its particularly generous contribution to resource surveys and education. Some 4,000 young Africans study in Romania and more than 12,000 specialists have been sent to African states. Despite their membership of the Warsaw Pact the Romanians have tried to gain access to the non-aligned movement in which Yugoslavia has for long had an influential position. Romania has been able to maintain and develop good relations with a range of developing countries, often in conflict with one another, and this indicates a degree of autonomy in Romanian foreign policy which other Warsaw Pact states cannot command. Romania even decided to retain links with Chile following the replacement of Allende's Marxist government in 1973, such was her interest in Chilean copper. Of course when pressures have become overwhelming Romanian compliance with Soviet policies has been assured and the increased level of Romanian economic dependence on the USSR, coupled with growing western disenchantment of the Romanian record on human rights, may reduce the scope for deviations in future policy. But certainly at the time of the Sino-Soviet dispute and the detente atmosphere of the 1970s the Romanians gained wide acceptance as an ally of the non-aligned movement and an active participant in United Nations' efforts to solve international conflicts.

NOTES

1.   M.R. Jackson (1976) 'Prices and efficiency in Romanian foreign trade', in Brada (ed.) 117-33.
2.   Bolitho (1971) 42. See also Joseph (1987).
3.   Bolitho (1971) 134.
4.   Hill (1983); Masnata (1974); Saunders (1980); Vine (1987); Wilczynski (1969). See also P.A. Boot (1987) 'East-West trade and industrial policy: the case of the GDR', Soviet Studies 39: 651-71.
5.   Jackson and Woodson (1984).
6.   J.C. Brada et al. (1982) 'The optimal rate of industrialisation in developed and developing centrally-planned economies: a general equilibrium approach', in C.K. Wilber and K.P. Jameson (eds) Socialist Models of Development, Oxford: Pergamon, 991-1004 (quote p. 1002). See also R.L. Farlow (1971) 'Romanian foreign policy: a case

of partial alignment', Problems of Communism 20(6): 54-63;
R. Forrest (1982) 'Romanian-American economic relations
1947-1975', in S. Fischer-Galati et al. (eds) Rumania
between East and West: Historical Essays in Memory of C.C.
Giurescu, Boulder, Col: East European Monographs, 385-413.

7.   R.O. Blalack (1972) The Role of Czechoslovakia in
Non-Comecon Trade, Georgia State University: Ph.D. thesis.

8.   John (1975). See also C.F.G. Ransom (1971) 'The
future of EEC-Comecon relations', World Today 27:438-48.
See also A.A. Duymovic (1980) 'Poland's agricultural policies
in the 1970s: impact of agricultural trade with the US', in
R.A. Francisco et al. (eds) Agricultural Policies in the USSR
and Eastern Europe, Boulder, Col: Westview, 185-98.

9.   Bertsch and McIntyre (1983). See also A.
Dezenyi-Gueullette (1983) 'The utilisation and assimilation
in Hungary of advanced technology imported from the West',
Soviet Studies 35:196-207.

10.   Haydon (1976); Levcik and Stankovsky (1979);
Paliwoda (1981); Wasowski (1970). See also P. Hanson (1978)
'East-West industrial cooperation agreements', in Shlaim and
Yannopoulos (eds) 127-59; E.A. Hewitt (1975) 'The
economics of East European technology imports from the
West', American Economic Review 65:377-82; U. Kivikari
(1988) 'Special operations: solutions or substitutes in East-
West trade', Nordic Journal of Soviet and East European
Studies 5: 55-62; I. Spigler (1975) 'Direct western investment
in Eastern Europe', St Anthony's College, Oxford, Papers in
East European Economics.

11.   Saunders (1981).

12.   J.M. Montias (1985) 'Observations on trends in
East-West trade in the 1970s', in Bethlen and Volgyes (eds)
129-40; Zloch-Christy (1988). See also M. Babic and E.
Primoran (1986). 'Some causes of the growth of the Yugoslav
external debt', Soviet Studies 38:69-88; R. Portes (1977)
'East Europe's debt to the West', Foreign Affairs 55:751-82.

13.   Coker (1985); Kanet (1988); Radu (1981); Stokke
(1968).

14.   C.W. Lawson (1983) 'National independence and
reciprocal advantages: the political economy of Romania-
South relations', Soviet Studies 35:362-75.

Chapter Three

TRANSPORT SERVICES BY RAIL AND WATER

The 1970s brought a great increase in the demand for transport. This was very evident in terms of passengers because the growth of the cities meant an increase in the number of journeys to work by public transport, while the intensification of business contacts is reflected in the growth of air and rail travel (Table 3.1).

However, it is the movement of freight which has been particularly significant because the increase in the scale of industry means that a greater effort is required to assemble raw materials and intermediates and to distribute finished products; while the increasing dependence on imports of energy and raw materials creates further demands which can be particularly heavy where the key factories are remote from the ports of entry into each country. The growth can be seen in the volume of shipping including a greatly increased merchant marine (5.5 million tonnes in 1971 and 8.1 in 1981 excluding Albania), the pipeline and rail flows between the Soviet Union and Eastern Europe and the rail traffic within Eastern Europe (Table 3.2). The volume of freight is bound to increase further. This chapter therefore examines the ways in which increases in capacity have been provided, with particular reference to railways and waterways. Investment choices for the future will involve some difficult questions including the further restructuring of the railways (through entirely new lines or modernization of existing links) and the exploitation of the waterways, calling for very heavy investment now against the prospect of benefits over the longer term.

Table 3.1 Transport communications and the media 1970-80

Country	Values per thousand of the population for													
	Motor vehicles		Merchant shipping (a)		Civil aviation (b)		Radio receivers		Television sets		Telephones		Newspapers distributed	
	1970	1980	1971	1981	1970	1980	1970	1980	1970	1980	1970	1980	1970	1980
Bulgaria	na	na	82	134	565	1,430	270	314(c)	121	186	56	141	193	234
Czechoslovakia	71	171	6	12	872	1,616	269	289	216	280	140	206	252	304
GDR	91	179	60	91	na	na	351	370	264	342	122	189	445	517
Hungary	23	110	na	7	50	162	244	252	171	258	80	118	214	242
Poland	23	68	54	100	99	351	174	295	130	224	57	95	209	237
Romania	na	na	18	91	62	197	151	144	73	167	31	56	169	181
Yugoslavia	41	95	76	114	107	348	166	207	88	192	36	95	85	103

Source: US yearbooks

Notes: (a) tonnage
(b) kilometres flown
(c) 1977

Table 3.2 Railway networks and traffic 1980

Country	Network			Density		Freight (billion tonne/km)				Passengers (billion passenger/km)			
	A	B	C	D	E	F	G	H	J	F	G	H	J
Albania	0.25	0.00	0.0	8.82	0.98	17.7	27.3	4.37	1.99	8.0	9.6	1.99	0.90
Bulgaria	4.04	1.65	40.8	36.47	4.55	66.2	18.4	5.04	4.33	18.0	-4.8	1.37	1.18
Czechoslovakia	13.13	3.03	23.1	102.67	8.59	56.4	35.9	3.97	3.37	23.1	30.5	1.62	1.38
GDR	14.22	1.68	11.8	131.43	8.50	24.4	23.2	3.20	2.28	14.7	-9.8	1.92	1.37
Hungary	7.62	1.55	20.3	81.80	7.11	134.7	35.6	5.53	3.75	51.3	39.0	2.11	1.43
Poland	24.36	7.09	29.1	77.89	6.78	75.5	57.3	6.80	3.40	23.2	30.3	2.09	1.05
Romania	11.11	2.37	21.3	46.78	5.00	25.0	29.5	2.66	1.11	10.3	-5.5	1.09	0.46
Yugoslavia	9.39	3.32	35.4	36.72	4.19								
Eastern Europe	84.12	20.69	24.6	65.99	6.24	399.9	34.3	4.75	2.97	148.6	18.1	1.77	1.10

Source: Statistical yearbooks.

Notes: A Railway routes (thousand km)
 B Railway routes electrified (thousand km)
 C Percentage electrified
 D Km railway per thousand km^2 of territory
 E Km railway per thousand inhabitants
 F Total freight/passenger movement
 G Percentage change over 1970
 H Traffic per kilometre of railway : million tonne/km or million passenger/km
 J Traffic per capita : thousand tonne/km or thousand passenger/km

RESTRUCTURING AND MODERNIZATION OF THE RAILWAY SYSTEM

In each country the railway network is of prime importance for the domestic economy, but the railways are also very important for international trade. Important routes link the East European capitals by way of Berlin and Prague through Budapest to Belgrade, Bucharest, and Sofia, or from Warsaw to the Balkans through Lvov (Soviet Union) to Bucharest and Constanta, whose sea link with Istanbul enables this route to be used as an alternative to the direct routes from western Europe to Turkey. Tourist traffic to the Black Sea resorts from Czechoslovakia, the GDR, and Poland is heavy and, along with the Adriatic Coast of Yugoslavia, justifies augmented services from western Europe also, on the routes through Vienna and Trieste. Freight movements by rail within Eastern Europe concern timber, fruit, and other raw materials from the Balkans in exchange for manufactured goods; Czechoslovakia and Hungary both play a key role in this international traffic by virtue of their central positions.

Albania's railways have been built from scratch since 1945 but in all the other countries considerable overhaul has been necessary to accommodate new traffic flows (and in some cases to adjust to new political frontiers). Increases in capacity on main lines through doubling (and in some cases the realigning) of track, dieselization, and electrification have occurred everywhere, and some lightly used branches have been closed, especially where a change of gauge inhibits integration with the rest of the network. (1) The building of underground railways may allow for some surface railways to be dispensed with. Thus in Budapest the closure of that section of the Godollo railway which is duplicated by the underground will provide space for the reconstruction of the Kerepesi highway as a dual carriageway and lead to the elimination of a serious road transport bottleneck. In general however the construction of entirely new railways has been very limited in both Hungary and Czechoslovakia. (2) The main project in Czechoslovakia has been the broad (Russian) gauge track from the Czechoslovak-Soviet frontier to the iron and steel works of Kosice. But reference may also be made to the rack-and-pinion railway running through the Tatra for 5 km between Strba (895m) to Strbske Pleso (Strba Lake) at 1,355m. The line was worked between 1896 and 1932 before being reopened using a 1 metre gauge in

1972. The funicular railway on Prague's Petrin Hill, built in 1891 but closed by landslip damage in 1965, was reopened in 1985.

The GDR

The reorganization of the railway system has been particularly important for Germany. Substantial frontier changes not only reduced the route length under German administration but also divided the smaller system between two managements: Deutsche Bundesbahn and Deutsche Reichsbahn. (3) The latter deals with the GDR and West Berlin. Also some lack of harmony has arisen in certain localities between the new frontier lines and the inherited rail network (Figure 3.1). Thus the railways on the island of Usedom were cut off from the rest of the German system because the junction at Swinemunde (Swinoujscie) passed under Polish control. Some special operating arrangements have been made, as in the case of the freight line from Gorlitz to Zittau which runs along the Polish side of the Oder and which is retained as a through route for the Germans by agreement with the Polish railways who work the German trains through their territory. But the special operating rights in the Asch (As) and Eger (Cheb) areas of Czechoslovakia have been withdrawn and local lines which formerly crossed the frontier have, with the exception of Varnsdorf, been closed or restricted to freight traffic. The division of Germany has resulted in the closure of most lines interrupted by the interzonal border but some improvement has followed from the Grundvertrag of 1972: out of a total of thirty-eight lines cut by the frontier nine remain in use (two of them solely for freight traffic). Lines which looped into the territory of the other German state have been seriously affected too. There is a particularly outstanding case involving some 15 km of track between Fortha and Gerstungen where the old route, which crossed the interzonal border, has been replaced, at the cost of very heavy earthworks, with a new line which stays in the GDR territory. In the FRG the branch to Tettau in Bavaria has been closed because of the intervening frontier and special road vehicles now carry the railway wagons from the main line station. Radical changes were made in Berlin to restrict the railways through West Berlin to a single main line. All domestic GDR traffic was diverted after the completion, in

Figure 3.1 Railways in the GDR and Poland

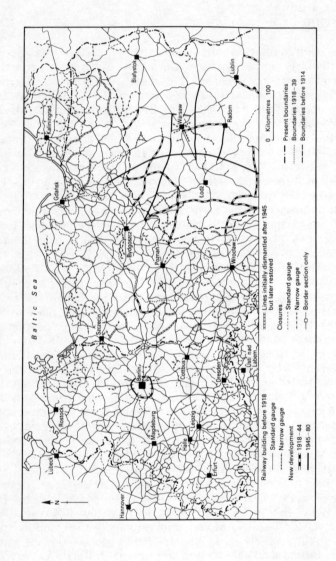

Source: Railway timetables.

1957, of a partially realigned Berliner Aussenring (BAR), including a series of complex junctions where the main radial lines from city centre stations cross it. There has been little other new construction apart from short branches in the Lusatian lignite field, but several rail (and road) diversions have been necessary in order to allow the extension of lignite quarries. There are several examples near Bitterfeld, including the railways to Leipzig and Wittenberg.

Formal changes in the network tend to obscure much of the post-war transformation. Traffic flows have changed out of all recognition. New international frontiers have greatly reduced traffic on lines that were formerly important domestic routes. On the other hand the GDR's understandable preference for its own ports of Wismar, and especially Rostock, has given rise to a major northward freight flow from the Halle-Leipzig area. Again the military interests of the Soviet Union have given priority to the line across the southern part of the GDR from Thuringia to the Polish frontier by way of Karl Marx Stadt, Dresden, and Görlitz: it continues eastwards through Poland and into the Ukraine. Lightly used lines have been closed altogether in some cases while the proportion of route length with double track fell initially from more than one-third to just 10 per cent, covering those lines with strategic importance for the Red Army. Double track was restored on the major freight routes in the 1950s and replacement has continued in the 1970s, but a prominent feature of the railways of the GDR is the modernization of routes to allow increased capacity in the context of single-track working. Capacity has increased through better signalling, higher average speeds, and heavier trains, while some heavy traffics are handled by diversions: thus trains from Berlin to Rostock may use the double-track line through Wittenberg and Schwerin as well as the direct line (which is single track north of Neustrelitz) rebuilt since 1945. Some railway electrification equipment was removed to provide reparations to the Soviet Union but this left the fleet of steam locomotives dependent on lignite. Low calorific value made steam raising difficult and high ash content created problems of waste disposal (before 1945 the railways had been able to use locomotive coal from the Ruhr and Upper Silesia). However steam has now given way to diesel traction, using both locomotives and railcar sets, although the necessary oil has of course to be imported. And electrification has started again, first with the freight routes comprising the Saxon Ring (starting in 1953). Work is now proceeding on the lines from Halle-Leipzig to Berlin and the Czechoslovak frontier at Decin. The Rubelandbahn has also been electrified to cope with the heavy mineral traffic between Blankenburg and Konigshutte in the Harz Mountains.

The Balkans

In the Balkans Bulgaria has concentrated on improved access to the ports of Burgas and Varna from Sofia. (4) In Romania a new axial line from Tecuci in Moldavia to Bucharest and Craiova has been completed while comparatively small projects north of Tirgu Jiu and south of Viseu de Sus have given more direct access to the Jiu Valley coalfield and the northern border region of Maramures respectively. (5) After the communist revolution the growth of the network slowed down initially because of the concentration of resources on the Danube-Black Sea Canal. However in recent years a number of branch lines have been built to open up the lignite fields (especially in Oltenia) while the rising cost of oil has brought a switch in priorities between road and rail during the 1980s, leading to the provision of rail links with towns like Tirgu Neamt and Siret and the shortening of rail distances between Bucharest and Transylvania through new construction between Curtea de Arges and Rimnicu Vilcea and between Brad and Deva. A number of projects however involve not so much the building of entirely new railways as the conversion of narrow-gauge lines to standard gauge and the realignment of railways which are technically inadequate for modern traffic requirements (Figure 3.2). Yugoslavia has completed several new railways such as Banja Luka-Tuzla, Bihac-Knin, Pula-Rijeka, and Samac-Sarajevo but the major development has been the Belgrade-Bar main line, a long-standing project dating back to 1920. (6) Along with the modernized Sarajevo-Ploce (Kardeljervo) railway this extends standard gauge into what was formerly the domain of the narrow-gauge military railway system initiated by the Habsburg Empire during the occupation of Bosnia-Hercegovina (Figure 3.3).

Albania's railway network is entirely a post-war development. (7) By 1950 lines from the port of Durres had reached Tirana and Elbasan. There have been three major extensions from this core system. In 1968 construction was

Figure 3.2 Railway system development in Romania

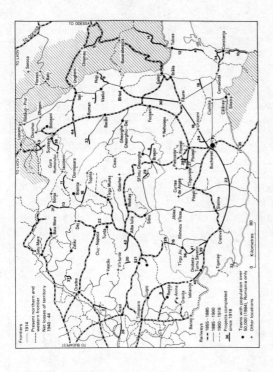

Source: D. Turnock (1987) 'The development of the Romanian railway network since the First World War', Leicester University Geography Department Occasional Paper 16. See p. 31 of this work for details of the projects numbered on the map. Projects 1-9 were completed between 1921 and 1938; 10-29 between 1939 and 1948; 30-37 between 1949 and 1960; 38-44 since 1960. Projects 8, 28, 29, 40, and 42 involved widening or diversion.

Figure 3.3 Railway system development in Yugoslavia

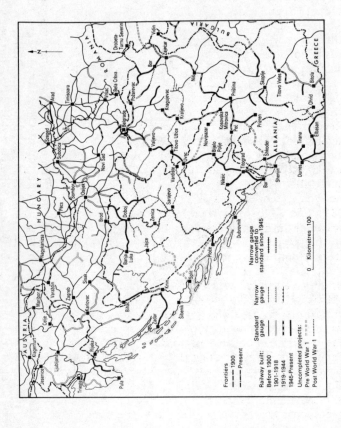

Source: Railway timetables.

completed from Peqin on the Elbasan line to Fier, with connections leading on to Ballsh and Vlore in 1974 and 1983 respectively. Then the Elbasan line was extended to Prrenjes in 1974 and Pogradec in 1980. Finally a line was built northwards from Vore, between Durres and Tirana, to Lac in 1962. This line has continued to Shkoder (1981). There may well be further railways built: from Pogradec to Korce and from Lezhe on the northern line to Klos and Shengjin, since the Albanians consider that rail transport is three times cheaper than road. The use of brigades of 'volunteers' mobilized from all parts of the country has made the railway projects important as symbols of national unity. Young people brought together in this context are encouraged to inter-marry and so help to overcome deeply held local and regional loyalties. The young people are also used to 'revolutionize' the population in areas where construction is being undertaken. However the future of international links is clouded by difficult relations with Yugoslavia. Polemics over the status of Albanians in Kosovo have helped to discourage the Yugoslavs from building a railway from Titograd to the Albanian frontier without which the Albanian system remains isolated. The Shkoder line was extended to the Yugoslav frontier (Han i Hotit) in 1985 and Albania's railways sought contact with the European system. Albania became a member of the International Union of Railways (1985) and joined the international convention governing reciprocal use of wagons in international traffic (RIV) in 1986. But because of the strained relations with Yugoslavia no through services are operating. Even the existing road transport links have been compromised. Albanian lorries have even lost their 'Most-Favoured-Nation' status on Yugoslav roads with the result that a thrice-monthly ferry service has operated between Durres and Trieste since 1983 in order to bypass hostile territory. A ferry link with Otranto in Italy is also planned; relations with Greece, first normalized after the war in 1971, have improved to the point where local co-operation agreements are being contemplated including a ferry service from Sarande to Corfu.

Despite all the new construction since 1945, the Balkan railway network will almost certainly never achieve the density of the northern countries because of the superiority of road transport for handling the relatively light traffic generated in rural areas. Furthermore increased capacity on inter-city routes can be provided by doubling track and electrification rather than by building entirely separate lines. In Bulgaria the line from Sofia to Mezdra was widened in 1969 while electrification gained momentum during the same decade: Gorna Oriakhovitsa to Ruse and Sofia to Plovdiv in 1963 followed by Sofia to Karlovo in 1968. Of the Bulgarian system 40.8 per cent is now electrified (30.3 in Romania and 35.4 in Yugoslavia).

Poland

It is however in Poland where major new projects have been embarked upon in recent years. (8) Increased flows of coal northwards from Silesia and of iron ore into Silesia from the Soviet Union have called for additional capacity. These requirements are additional to the transit traffic between the USSR and GDR which led to the Mazowiecki-Radom and Skierniewice-Lukow connections in 1949 and 1953 respectively and other lines such as Wloszczowice-Tarnobrzeg (1973) and Piotrkow-Belchatow (1977) geared to local traffic. The current projects are known as the Linja Hutniczo-Siarkowa (LHS), 'the steel and sulphur line' and the Centrala Magistrala Kolejowa (CMK), the central trunk. It was in 1971 that work began on a new standard-gauge main line from Upper Silesia to the Soviet frontier at Hrubieszow/Vladimir Volinski via Kielce; 70 km (Zwierzyniec-Nisko) were complete in 1976 when the project was altered to a broad-gauge electrified line that would eliminate the need for trans-shipment at the frontier. For 175 km the line runs parallel to existing standard-gauge track but the remaining 225 km use a completely new alignment, with no intermediate stations. The LHS was seen as the largest transport investment in Polish history, involving the use of heavy rails (never previously used in Poland) and an advanced signalling system, to handle trains of 4,800 tonnes maximum. However the work appears to be incomplete and there are reasons for supposing that the investment was fundamentally misconstrued (leaving aside the country's present economic crisis). The line is not helpful for passenger traffic and freight operations are complicated by trans-shipment in Silesia. Ore for Huta Katowice will be taken by conveyor from the stockpile at Slawkov (where the broad-gauge railway terminates) and any ore for Nova Huta routed by the LHS will require trans-shipment followed by a standard-gauge rail haul in the eastward direction! Finally electrification has limited value since the Soviet line from the frontier to Kiev and Kursk is not yet electrified. It is reported that part of the LHS is open for preliminary traffic and that in the meantime the Poles are falling back on the standard-gauge line from Krakow to Przemysl.

Poland had also given much thought to the improvement of railways to the Baltic ports. In the case of Szczecin-Swinoujscie the duplication of track and electrification of the line from Opole via Wroclaw and Zielona Gora is an obvious strategy, providing a link with annual capacity for coal shipments of 40 million tonnes. Similar work could proceed in the case of the railway from Katowice to Bydgoszcz and Gdynia but potential has been considered great enough to justify the new high-speed CMK from Katowice to Grodzisk and Warsaw with an extension from Grodzisk to the Baltic ports. The CMK would appear to be a much more promising project providing not only much-needed capacity for domestic traffic between Upper Silesia, Warsaw and Gdansk-Gdynia but also an efficient link between the expanding ports and an East European hinterland embracing Czechoslovakia (via the Zebrzydowice/Petrovice frontier crossing) as well as Poland. The modernized railway from Szczecin to Opole can also extend across the frontier at Chalupki/Bohumin and into Czechoslovakia. Both lines feature in the long-term programme for railways with improved equipment permitting heavier trains and faster running. It is reported that a single-track railway now exists from Zawiercie to Grodzisk (233 km). It is not yet electrified but is used by some freight and passenger traffic in 1984.

Future prospects

Railway modernization is continuing with the prospect of increasing importance due to the energy crisis and the need to reduce road transport where possible. Work includes improvements to signalling, track, rolling stock, and station facilities. There has been an increase in the scope of express trains now that much of the local transport is handled by the roads. The improvements in Romania between 1948 and 1977 are shown in Figure 3.4. However it is also noticeable that the emphasis on petrol and diesel railcars which was growing in the 1930s has been somewhat reduced owing to the

Figure 3.4 Railway modernization in Romania

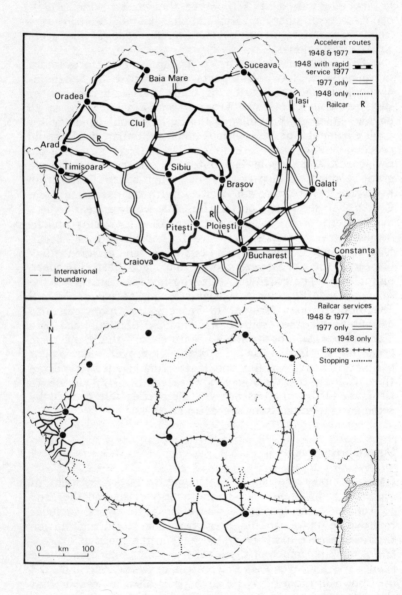

Source: Railway timetables.

changes in energy policy after 1948 (with priority for exports to the Soviet Union and the increases in supply to the domestic chemical industry) and reduced interest in high-speed rail travel (a function of the growth of domestic air services as well as the higher priority for freight on the railways). However there have been gradual improvements in train speeds, particularly noticeable in the south-east of the country and also in Transylvania where changes in motive power on steeply graded lines have been particularly beneficial (Figure 3.5). Electrification now affects a quarter of the railways in Eastern Europe and the process is accelerating on account of the critical shortage of liquid fuels and decreased production of diesel locomotives. However the benefits of the investment are not always as great as they could be because of the neglect of parallel improvements in track and signalling without which increases in speed and capacity cannot be obtained.

International co-ordination of electrification has been limited, although modernization work agreed in principle in 1979 involves reconstruction of certain main lines, especially three lines originating in the USSR and serving Poland and the GDR (Dorohusk-Ostrow-Tuplice-Forst-Falkenberg); Hungary (Batevo-Zahony-Budapest), and Romania/Bulgaria (Ungheni-Bucharest-Giurgiu-Ruse-Gorna Oriakhovitsa-Sofia). Also included in the scheme are lines running south from the Baltic ports of Gdynia (to Zebrzydowice, Zilina, Budapest, Biharkeresztes, Brasov, and Bucharest) and Rostock (to Berlin, Bad Schandau, Prague, Sturovo, Szob, and Budapest). The aim is to raise maximum train capacities to between 3,000 and 4,800 tonnes and maximum speeds to between 100 and 140 km per hour. When it is considered that express passenger train speeds on some main lines are no more than 75 km per hour these improvements appear substantial. Increased use of interchangeable bogies would be involved on these lines but for traffic that will still have to be trans-shipped there is an intention to make more use of containers. These were first used in 1967 when an agreement between Russian and Swiss forwarding companies opened up the 'Siberian Land Bridge' between Japan and western Europe. The traffic soon built up into whole train loads, passing through Eastern Europe in transit, because the land route proved to be much quicker than the sea voyage (one month as against two) and also cheaper, on account of competitive freight rates and savings on insurance and packaging. (9)

Figure 3.5 Improvements in railway passenger train speeds in Romania

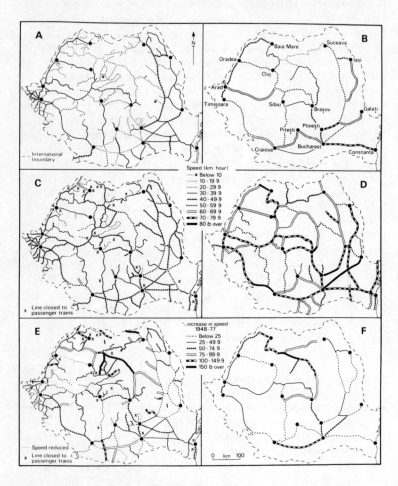

Source: Railway timetables.

Notes:
A. Stopping trains in 1948
B. Express trains in 1948
C. Stopping trains in 1977
D. Express trains in 1977
E. Improvements for stopping trains 1948-77
F. Improvements for express trains 1948-77

Czechoslovakia began operating container trains to the GDR in 1971, the USSR in 1973, and Poland in 1974. Some ten container terminals were reported in use in 1976, the first being at Bratislava, Brno, Decin, Plzen, Prague, and Prerov. Czechoslovakia's interest in containers is quite understandable. As an inland country there is a great interest in reducing costs while the central position in Europe gives good potential opportunities for development of transit traffic. Under a Comecon agreement of 1974 Czechoslovakia was made responsible for operations on four scheduled container train routes: Sofia-Bucharest-Budapest-Prague-Berlin-Rostock (the Decin-Rostock section was inaugurated in 1971 as the first regular container train in Eastern Europe); Budapest-Katowice-Warsaw-Gdynia; Plzen-Prague-Kosice-Kiev-Moscow; Prague-Wroclaw-Poznan-Szczecin. A special depot for the transfer of containers operates at the Polish-Soviet frontier (Brest/Malaszewicze). However the total amount of freight handled in 1975 was less than half a million tonnes and insignificant in relation to the total amount of railway freight (less than 0.1 per cent). Poor equipment at the terminals and a lack of road haulage units for containers were cited as constraints at that time.

PORT DEVELOPMENT

The East European countries possess modern fleets, even the land-locked states of Czechoslovakia and Hungary. Moreover a number of vessels are available for international charter. The growth has arisen out of the widening scope of commercial contacts across the world geared to the support of allied states like Cuba and Vietnam and the growing trade in certain key commodities (like cereals, oil, and wool) with capitalist states. Ships have become much larger, following developments in the rest of the world, and maximum utilization of these vessels has led to the opening up of various 'triangular' shipping routes. Further flexibility has been achieved through the construction of vessels that can carry both dry and liquid cargoes. The GDR and Poland have developed their Baltic ports not only for their own requirements but also to attract traffic from Czechoslovakia and Hungary. The GDR's development of Rostock under difficult conditions complements the Polish effort at Szczecin and Gdansk-Gdynia. Yugoslavia's

101

principal Adriatic ports are Rijeka and Split which, along with Trieste in Italy, handle some of the transit traffic to and from the land-locked states. Additional capacity is being installed at Bar and Kardeljevo backed up by better rail links across the Dinaric Alps already referred to.

North Sea: Poland and the GDR

Although Poland gained access to a 500-km-long coastline after 1945, compared with only 73 km before, the low volume of international trade along with the emphasis on exchanges with other socialist countries prevented major investments in the ports until the 1970s. (10) There are two major port complexes: Gdansk-Gdynia and Szczecin-Swinoujscie. There is also the minor port of Kolobrzeg and a number of fishing ports. Between 1950 and 1980 the number of vessels using Gdansk-Gdynia and Szczecin-Swinoujscie went up from 4,638 to 4,907 and from 3,618 to 5,634 respectively. Increases in total tonnage were much greater of course: from 5.11 million to 18.87 and from 2.26 to 11.55 respectively. Comparable figures for cargo throughput are 10.13 to 36.24 and 5.19 to 24.74. Passenger traffic has increased from a negligible level in 1950 to 302,000 at the two complexes combined. Although coal and coke were easily the most important commodities handled in both 1970 and 1980 (16.5 and 21.1 million tonnes respectively, representing 45.5 and 34.4 per cent respectively of all cargo handled) the greatest increases during the decade occurred in ores (3.0 to 7.1 million tonnes), grain (1.5 to 8.7), and petroleum (2.0 to 7.3). However iron ore goes mainly to Szczecin, while grain and timber are handled mainly at Gdynia and Gdansk respectively. Both are important for raw materials related to fertilizer manufacture which takes place at Gdansk and also at Police near Szczecin. Passenger traffic is also considerable in view of the regular ferry services to Scandinavia (Gdansk to Helsinki and Swinoujscie to Ystad) which form part of the north-south transit routes through Eastern Europe. Gdynia is a port of call for cruise liners and offers a service to Montreal using the 15,000 tonne vessel 'Stefan Bathory'.

Wartime damage in the ports has been repaired and major extensions have been made to cater for larger vessels. Work at the new port of Gdansk began in 1970 and facilities for coal, oil, and iron ore opened in 1974, 1975, and 1978 respectively; the new northern harbour (Port Polnocny) can accommodate ships of up to 130,000 dwt and offers a pipeline link with Plock. Since Gdansk has its own refinery, an important function of the pipeline is to send crude oil into the Friendship pipeline for transport on to the GDR. The Gdansk project complements the four phases of development at Swinoport (Swinoujscie) which culminated in the construction of a new coal and ore terminal in 1983. Swinoujscie has special facilities for handling chemicals, coal, grain, and iron ore as well as general cargo. Szczecin has a container base (at Zdunow) and like Gdynia specializes in grain and containerized general cargo.

The hinterland of the Polish ports extends to other socialist countries and pressure for better transit facilities for Czechoslovakia and Hungary (not to mention Austria, Romania, and Yugoslavia which all have an interest in the Baltic ports) led to the setting up of the Interport organization. Czechoslovakia maintains its own merchant marine, based at Szczecin. The ships carry Czech goods to various countries in Asia and also contribute to the work of the Cubalco Line (linking Europe and Cuba) formed in 1963 by Czechoslovakia, the GDR, Poland, and Cuba. There is therefore a significant Czech community in Szczecin with its own housing units and schooling facilities. The GDR ports are also involved in Interport. Rostock is the chief port and its 'political clout' as the chief town of the coastal region no doubt ensured its selection for rapid expansion over Stralsund and Wismar. Rostock, along with Sassnitz, has a ferry connection with Klaipeda in the USSR using new purpose-built vessels with two broad decks and simultaneous loading. However the formation of Interport has led to increased GDR interest in Szczecin. Until the 1970s there were strained relations as the GDR tried to frustrate Polish investments at Szczecin in the hope that the port and its immediate hinterland might be transferred to the GDR. (11) In the immediate post-war years there were also difficulties for the Poles arising out of the strong Soviet presence at Szczecin in connection with the shipment of reparations from Germany. But Silesian coal was being exported during the 1950s, using an elaborate system of conveyor belts, and the channel was deepened to 5 fathoms, suitable for ships of 10,000 tonnes dwt.

The Balkans

The ports of the Balkans serve their respective countries but there is some use of them by Czechoslovakia and Hungary while Bulgaria and Yugoslavia despatch traffic through the Greek port of Thessaloniki. (12) There is an interesting link with river traffic on the Danube through the Interlighter international shipping enterprise jointly created by the Soviet Union, Bulgaria, Czechoslovakia, and Hungary. The idea is to integrate river and sea transport by introducing container barges (of 1,100 tonnes) which can be transferred bodily to sea-going vessels (of 36,000 tonnes) at the rivermouth. The company is expected to handle about 1.5 million tonnes of goods annually. The first container ships 'Juliuc Fucik' and 'Tibor Szamuely' each accommodate 26 lighters and maintain a maximum speed of 19 knots. Routes have been opened up to Interlighter deep-sea sailings run from Ust-Dunajsk (Vilkovo) in the Danube delta's northernmost Chilia/Kilia channel to Pakistan (Karachi), India (Bombay), Sri Lanka (Colombo), Malaya (Penang), Indonesia (Port Kiang), and Vietnam (Ho Chi Minh/Saigon). The lighters are unpowered barges and require the assistance of tugs. But they are all-purpose vessels and can be used for general consignments as well as for bulky loads. Containers can also be handled. The service is well suited to trade with developing countries with poor port facilities. It is used not only by member states but also by other Danubian states: Austria, FRG, Romania, and Yugoslavia.

Another recent development is the opening of a train ferry service in 1978 between the Bulgarian port of Varna and the Soviet Union's general cargo port of Ilichevsk near Odessa. Four 22,000 tonne vessels carry 108 freight cars and operate the route on a two-day cycle. So far traffic has not exceeded 3 million tonnes per annum: less than one tenth the total freight traffic between the Soviet Union and Bulgaria. Problems have been experienced over the marshalling of wagons at Ilichevsk and the loading of the ferries, since wagons have to be moved between docks on elevators. Traffic should grow to 5.5 million tonnes by the end of the decade when another four ferries of modified design join the fleet. At present the ferry carries Soviet metals, oil products, fertilizers, building materials, timber and machinery and Bulgarian foodstuffs, chemicals, and equipment. The ferry avoids the rail haul across Romania (a long journey since the line has not been upgraded) and avoids the need for trans-shipment since broad-gauge wagons can operate in Varna which functionally may be regarded as an enclave of Soviet heavy industry. The special freight cars (comprising a pool of 1,500) can operate elsewhere in Bulgaria with a bogie change at Varna.

INLAND WATERWAYS

Inland waterways would appear to constitute an energy-efficient transport system but their use is very limited. Thus only 3.0 per cent of Czechoslovakia's freight (tonne-kilometres) was handled by waterways in 1975 and the proportion in 1990 is estimated at only 4.8 (with 9.7 for pipelines and the rest allocated to rail and road). Inland waterway traffic in Eastern Europe highlights a limited number of major rivers like the Oder and Vistula. Navigation often requires a measure of international co-operation. The Elbe was formerly important as an outlet for Czech goods moving northwards for export via Hamburg, but the incidence of the German interzonal frontier between Hamburg and Magdeburg has encouraged some transfer of traffic to the railways to Rostock or Szczecin. (13) The low terrain of the North European Plain reduces the cost of building link canals in places where the natural river systems do not coincide with the main lines of trading movements. Canals were built to link the Rhine with the Elbe, Oder, Vistula, and Dnieper but, except in the case of the larger Mittelland Canal in Germany, their capacity is too small for them to be valuable today. To accommodate larger vessels much higher levels of capital investment are required. However, the potential traffic is being constantly reduced because the inherent slowness of water transport often makes rail or even road transport preferable.

The Danube

The Danube is certainly the most important river. Limited traffic potential and lack of unified political control in the past often discouraged the necessary stimulus to overcome the main physical hazards. The removal of the Turkish economic monopoly in the nineteenth century and the expansion of the cereal and timber traffic eventually prompted improvements at the critical sections of the Iron

Gates, where the Sip Canal was built (through which vessels moved upstream against the strong current with the aid of a steam locomotive running on the western bank), and the delta where the Sulina river was dredged, straightened, and dyked at the outfall. The inter-war political situation was hardly conducive to further development but greater co-operation between the riparian states, including Austria and the FRG, has been achieved since 1945 and improvements to the river have been made still more feasible by a combination of navigational, hydro-electrical, and water management interests. (14) The production of electricity at the Iron Gates, where twin power stations have a combined capacity of 2,000 MW, was crucial in enabling Romania and Yugoslavia to go ahead with a massive joint project which has finally overcome the navigational hazards of this difficult section. (15) Further co-operation with Bulgaria at Turnu Magurele and with Yugoslavia at Gruia is following this pattern. Underpinning the whole Danube strategy is a desire for greater flood control, particularly important now that most of the natural flood plain has been dyked and drained for intensive cultivation. At the same time the major Danube tributaries are being harnessed and this will allow some extensions to the navigable waterway system, possibly with respect to the Olt and Siret.

The River Danube is now used for a variety of cargoes. Soviet raw materials including coal, forest products, iron ore, and oil move upstream, much of the traffic coming by barge from Ismail and Reni where trans-shipment from sea-going vessels takes place. Iron and manganese ores, for example, originate from Black Sea ports like Nikolaev and Kherson. The total traffic of Ismail and Reni has risen to some 12 million tonnes and includes substantial cargoes moving downstream for export to the Gulf States and Eastern Mediterranean. (16) Foreign trade via the Danube has also increased through the use of river-sea vessels. These were first introduced in the Soviet Union in 1962 and their use has been considerably expanded recently. The vessels of 1,000-3,000 tonnes capacity can operate on the main inland waterways and also in coastal waters, subject to certain specified constraints over distance and wind/wave strength. The system has been used for exchanges of foodstuffs and metals between Bulgaria and the Soviet Union. There are now regular consignments of timber (originating in the Komi ASSR) from the Soviet Union to Bulgaria.

Mention should also be made of traffic crossing the Danube. On the Lower Danube a road bridge has been built on Romanian territory between Giurgeni and Vadu Oii as part of the main road between Bucharest and Constanta and the historic railway bridge built by Angel Saligny to connect Fetesti and Cernavoda has been duplicated by a combined rail-road bridge. There are international links for rail between Giurgiu (Romania) and Ruse (Bulgaria) and for road at the Iron Gates dam between Romania and Yugoslavia. In the Middle Danube road and railway bridges suffered serious damage during the Second World War and some, like the railway bridge at Esztergom, have never reopened. There are however numerous bridges in use in the vicinity of three great cities: Belgrade, Bratislava, and Budapest. At Budapest alone there are six road bridges (including the historic chain bridge 'Lanchid Hid' built by Istvan Szechenyi in 1849) and two railway bridges, with a further road bridge 'Lagymanyos' under construction and three further bridges and a tunnel in prospect for the end of the century largely in conjunction with the proposed motorway ring. Outside the main conurbations there are road-rail crossings at Komarno-Komarom, Dunafoldvar-Solt, and Bataszek-Baja plus a rail link between Osijek and Sombor. There are also ferry services; principally in Romania, between Braila-Smirdan and Calarasi-Ostrov and internationally between Calafat in Romania and Vidin in Bulgaria. Passenger services also operate on certain sections of the river, principally along the three main distributaries in the Danube delta (Tulcea to Periprava, Sulina, and Sfintu Gheorghe) and from Orsova upstream to the Yugoslav frontier.

Romania: the Danube-Black Sea Canal

Romania can offer similar trans-shipment facilities at Braila and Galati but since 1984 she has been in a position to compete more effectively because the Danube-Black Sea Canal (Figure 3.6) gives access to Constanta, a port available for large ships which cannot enter the Danube delta on account of a 7.3 metre limit at the entrance to the Sulina channel (the new Constanta Sud harbour will be able to take vessels of up to 150,000 tonnes. (17) Goods coming into Constanta by sea can be transferred directly into barges whereas previously the railway haul to Cernavoda meant an additional break of bulk. The canal will be useful

Figure 3.6 The Danube–Black Sea Canal

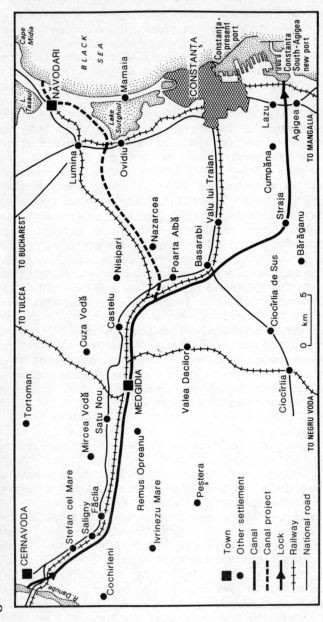

Source: D. Turnock (1986) 'The Danube–Black Sea Canal and its impact on southern Romania', GeoJournal 12:65–79.

to Romania for the movement of bulky raw materials destined for factories situated along the Danube. The present stage of development of the Romanian economy is conducive to a string of major industrial projects along the banks of the Danube and increased interaction by water to the point where a Bucharest-Danube Canal and a Danube-Black Sea Canal can be contemplated. The industrial projects tie up with the processing of imported raw materials with power drawn from the electricity grid: a generation ago the greater dependence on domestic raw materials and lack of a comprehensive grid system ruled out the Danube valley as an industrial axis. In addition to the economic justification which may become increasingly valid over the long term there is an important political factor: a great effort involving the whole nation and enhancing the image of the Communist party, which was forced to abandon the scheme when it was first implemented. Nevertheless the project has been very expensive, providing a capacity which the domestic economy cannot fully use in the foreseeable future.

Other countries may use the canal, which will be all the more attractive with increases in the size of Danube barges to 1,800 tonnes (likely if the Rhine-Main-Danube Canal is completed) since it can handle convoys of up to six 3,000-tonne barges. However, the Sulina channel through the Danube delta will be kept open and substantial investments have been made to extend the dykes out to sea and so protect the shipping lane from the expanding Chilia delta. (18) Ocean-going ships will continue to sail upstream from Sulina to the Romanian ports of Braila and Galati and the Russian ports of Ismail and Reni with transfer of cargo to river barges in many cases. Meanwhile Bulgaria has contemplated a canal from Ruse to Varna while Yugoslavia has shown considerable interest in developing the Danube system in order to reach the Adriatic coast (by the Sava) or to gain access to the Greek port of Thessaloniki through the Morava-Vardar corridor. (19) None of these schemes has yet been implemented.

Hungary and Czechoslovakia

Landlocked socialist states are particularly interested in shipping routes to link them with the oceans and it is notable that both Czechoslovakia and Hungary are playing their part in plans to improve the Danube. (20) Harbour installations have been completed at Bratislava and Komarno on the Czechoslovak side and both countries are involved in a project for shipping locks and a hydro-power station at Gabcikovo-Nagymaros. Ultimately there will be better navigation facilities and flood protection over the whole section of the river from the Austrian border to the Great Plain below Budapest, where Hungary plans locks and power stations on the Adony-Mohacs section.

The Gabcikovo-Nagymaros project was the subject of an agreement between Czechoslovakia and Hungary reached in 1958 with construction planned for the two hydro-electric stations for 1961-5. The work could not be financed at the time and a further agreement in 1973 envisaged construction to run from 1976 to 1985; but only be superseded by a further contract, modified in 1983, which puts back the completion date until 1995. The reasons for the repeated delays are not entirely clear but the scheme is technically complex and both parties have been heavily preoccupied with the development of nuclear power. Even the first two reactors at Paks offer more capacity than the complete hydro-power potential of the Danube and Tisza combined (880MW compared with 826). Hungary's share of the proposed Gabcikovo-Nagymaros works would be 443MW and a further 215MW could be installed in conjunction with schemes south of Budapest at Adony and Fajsz. The remaining 168MW arises out of 36 small projects, a number of which have already been completed. There is opposition to the scheme from conservationists while the uncertainty over the completion of the Rhine-Main-Danube Canal has reduced the urgency over provision of the minimum depth of 3.5m required for 1,350-tonne barges.

Working downstream the first part of the project is a dam at Dunakiliti which will impound a large reservoir supplying water to a new navigational channel running at a higher level than the present one as far as the Gabcikovo dam (while the old channel is retained to evacuate excess water). The flow in the new channel will be rapid (some 4,500 cusecs) and will be 14m above ground level, thus necessitating supporting walls 18m high. Barriers will control the rise of the water table on Grain Island which, as its name suggests, is an important granary. Drainage and irrigation have converted Grain Island into a prime area for production of cereals, fruit, and vegetables. Perishable produce is transported north to Ostrava and Prague where local conditions are not conducive to such a wide range of production. The well-rounded development of the area is brought out by the expansion of manufacturing (especially the shipyards at Komarno where ocean-going ships are now being made in addition to river boats) and tourism, expressed through the thermal lidos of Calovo, Dunajska Streda, and Patince. Gabcikovo power station will have a capacity of 700MW but will operate for only five hours each day when peak demand arises. For the rest of the time water (purified by the gravel strata) will accumulate in the Hrusov-Dunakiliti basin. Below Gabcikovo water will proceed to the turbines at Nagymaros where the capacity of 146MW will be exploited on a twenty-four-hour basis. Water will also be drawn off for irrigation and fish breeding. But, most important, in future it will be possible to navigate day and night and to use vessels of 1,600 tonnes pushed or towed by a new generation of motor boats.

The Hungarians have also been stimulated to give serious consideration to a Danube-Tisza Canal from Dunaharaszti (20 km south of Budapest) to a point on the Tisza between Csongrad and Szolnok. The 125 km canal would allow goods to move across Hungary from the Soviet to the Austrian frontier and on to western Europe. But any development of the Tisza route will require break of bulk at Zahony and so there is bound to be an important future role for the lower Danube through a more efficient integration of river shipping with Black Sea traffic. The Tisza is perhaps the most impressive example of improvements on a tributary of the Danube. From the turn of the century work has been going on to straighten the course of this tortuously meandering river and the present course is almost entirely artificial. Through shortening the course of streams in the Tisza basin (by several thousand kilometres in all) the gradient of the bed has been steepened and the rate of flow increased. S. Samogyi has calculated that the combined course of the Danube and Tisza (along with various tributaries like the Bodrog, Drava, Koros, Maros, Raba, and Szamos) has been reduced by 22.7 per cent from 3,835 km to 2,369 km as a result of a total of 594 cut-offs. (21) The total length of the cut-offs is not known but on the Bodrog, Drava, Koros (including the various headwaters), and Tisza 456 cut-offs with a total length of 431 km have replaced 1,657 km of meandering river.

Rivers are now more deeply incised and embankments provide additional security against flood but at the same

Figure 3.7 Canal projects related to the Danube

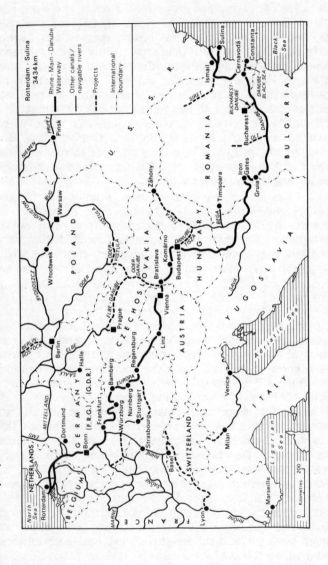

Source: D. Turnock (1979), 'The Europa Canal and its Central European Context', Bulletin
Society of University Cartographers 13(2): 32-6.

time flood periods have been shortened and the duration of low water has lengthened. Hence the need for dams to regulate flow and supply irrigation water. Dams on the Tisza have been at Tiszalok (1954) and Kiskore (1973) while further works are projected at Csongrad (and also in Yugoslavia) downstream and at Zahony and Vasarosnameny upstream. On the Koros tributary the first dam at Bokeny which dates back to 1905 was supplemented during the Second World War by Bekesszentandras (1942) and more recently by Bekes (1969). The dry land between the two rivers is irrigated by the Eastern Canal, running south from Tiszalok to the Koros (108 km) and will be supplemented by the projected Nagykunsag Canal associated with the Kiskore Dam. Agriculture has extended some 2 million hectares to former marsh and swampland and the old distinction between flood-plain and flood-free ground eliminated. Moreover the areas of sandy puszta have receded to little more than 100,000 ha (Bujac and Hortobagy). Housing conditions in some areas remain poor but the area is now dominated by prosperous co-operative farms and local industries like the shoe factory (with its related workshop turning out boxes) and sunflower oilpress noted recently at Tiszafoldvar near Szolnok.

Links with the Rhine, Oder, and Vistula

Also relevant to the Danube in Eastern Europe is the new canal being built to link the Rhine and Danube and to open up a waterway extending across Europe from Rotterdam to Sulina on the Black Sea. (22) The Rhine-Main-Danube Canal, suitable for the use of 1,350-tonne river barges which require a depth of at least 3.5 metres, will have considerable implications for Eastern Europe (Figure 3.7). The original Ludwigkanal between the Main and Danube built between 1836 and 1845 is now too small to be useful. An enlarged canal was envisaged as early as 1917 and subsequently included in the Nazi programme for the southern part of Grossdeutschland, for completion in 1945. Interest has been maintained because the scheme is multi-purpose involving some production of hydro-electricity and scope for both water supply and recreation. Clearly the real economic case to be made for these additional functions has an important bearing on the overall viability. For the navigational angle considerable advantage will accrue to socialist countries. Certain categories of freight may be transferred from rail to water and it may become feasible to build linking canals so that more industrial centres can be integrated with the water highway. In view of the time and cost involved in negotiating the various locks between Sulina and Rotterdam it is unlikely that the Danube-Rhine route will compete with ocean shipping between the Balkans and the Low Countries, but Romania clearly has an eye on the development of Danube traffic because a free port has been established at Sulina.

The impact of the canal is seen very positively in Hungary where the free port of Csepel (built in 1928) is set for expansion. With river traffic flowing in both directions (and scope for transfer to ocean-going ships at Reni and Ismail) and efficient transfer to road and rail transport at the dockside Csepel will be a key industrial location which it is to be hoped will attract foreign capital. At present the port handles a wide range of cargoes with iron ore and grain relatively prominent. Heavy investment is also being made in Czechoslovakia at Palenisko (Bratislava) and Komarno which will benefit not only from the development of the canal system but also from the Gabcikovo-Nagymaros regulation scheme which will allow larger vessels to be used and will reduce the fluctuation in water level from 8 metres (regularly experienced today through seasonal trends) to no more than 2 metres. Palenisko will be strongly bound up with oil traffic; the change to loading in a closed pool rather than the open river should reduce the danger of pollution. However, although only a small section of the canal has still to be completed the ultimate outcome remains very much in doubt. Present investment levels will not allow completion this century. Motivation is currently lacking on account of the poor economic prospects, coupled with opposition from conservationists and government concern in the FRG regarding Soviet claims (arising from the Belgrade Act of 1948, when the present Danube Commission was set up) to unrestricted use of the new waterway as far as Ulm. Even some of the main supporters of the canal, in Austria and Bavaria, are interested primarily in the waterway as a lever to use against the railways to gain lower rates.

If the Rhine-Main-Danube Canal is not completed there may be greater interest in Eastern Europe in linking together the Danube, Elbe, and Oder with new canals across Czechoslovakia, under a long-term Comecon co-operation agreement. An Oder-Danube Canal from Bratislava to

Raciborz with branch canals to the Elbe and Vistula could finally integrate the major rivers of Eastern Europe into a unified navigation system. (23) It was part of a Nazi strategy to establish Vienna as a counterpoise to Hamburg but the stress of war was too great for construction to proceed for long. The Oder-Danube project was reconsidered after the war in the context of Soviet strategy (and again in 1976), but the dream remains unfulfilled, although it would certainly offer some advantage to Czechoslovakia whose vessels work on the Oder from Kozle to Szczecin. Domestically there would be great advantage for Czechoslovakia in forging a waterway connection between Bratislava, Ostrava, and Prague. Czechoslovakia has allocated more investment to waterways since the late 1970s with the modernization of the Elbe waterway from Usti nad Labem to Kolin and the development of the lower sections of the Morava and Vah as important priorities. This work will allow for the distribution of coal to Bohemian power stations and the transport of oil products in Slovakia. But investment is insufficient to contemplate connections between the Danube, Elbe, and Oder during this decade.

The Oder has considerable potential for Poland (more than it previously had for Germany) and substantial improvements one or two decades ago would have been very beneficial in linking Upper Silesia with Szczecin. A Silesian Canal, linking the Oder and Vistula through the developing Rybnik coalfield, would expedite the flow of coal for domestic use, or for export through Szczecin-Swinoujscie. However the improvements carried out by Germany in the 1930s have now deteriorated and remedial works are beyond the limits of Poland's present financial capacities. Traffic has been diverted away from the river and it is used on a local basis only by boats of 500 tonnes. By the early 1980s less than 1 per cent of all freight traffic in Poland was being handled by the inland waterways, mainly sand and gravel travelling distances of about 100 km on average.

The Vistula is becoming a major industrial axis, as first envisaged in the plan of the 1930s for a Central Industrial Region at the Vistula-San confluence. The idea has persisted, leading to the preparation of a twenty-year Vistula Plan which was sanctioned in 1978. Demands for water supply, flood control, and hydro-electricity will require the construction of a string of reservoirs. Eventually there may be more than twenty power stations with a combined capacity of 2,000MW but at present Wloclawek,

115

with 160MW capacity opened in 1969, is the only significant producer: eight other dams already projected downstream of Warsaw would provide another 660MW. The navigation interest has not been adequately considered in the execution of the plan and this rules out the Oder-Vistula canal as a viable project. It also undermines any suggestion of a Vistula-Bug link across the Soviet frontier although the economic case is gravely weakened by the ample rail capacity available to move the iron ore. Some work was carried out in the years 1974-7 with a view to deployment of pusher combinations (up to 3,500 tonnes) on the Oder and Vistula by the end of the century, but investment could not be sustained.

The GDR meanwhile has some interest in the Elbe and Oder, but the estuarine ports are situated in other countries in both cases. National interests thus favour domestic schemes to improve access to Rostock, by a new canal from Berlin or else by a canal from the Elbe at Wittenberg, bearing in mind that Berlin has good links with the Elbe via the Ihle Canal (1866-71) which provided a more direct route than the one previously available through Magdeburg. The GDR has already built a 35 km canal to avoid West Berlin. The future therefore is distinctly problematical.

NOTES

1.   E. Barnes (1977) 'Motive power of the Czech railways', Railway World 38:384-7.

2.   I. Berenyi (1979) 'The Budapest Metro', Modern Railways 36: 259-62; Kalla-Bishop (1973).

3.   J.V. Gough (1980) 'Deutsche Reichsbahn today', Modern Railways 37:312-16; 365-9; J. Plang (1976) 'The state railways of East Germany', Railway World 37: 284-9; J.H. Price (1975) 'Berlin revisited', Railway Magazine 121:166-71. For narrow-gauge railways see D.E.J. Benn (1980) 'Reichsbahn narrow gauge', Railway Magazine 126:364-8; D. Eatwell (1985) 'Narrow gauge steam in East Germany', Railway World 46:18-25; O.W. Lautsen (1973) Bygone Light Railways of Europe, Lingfield: Oakwood Press, 104-8.

4.   P.J. Perry (1974) 'Bulgarian bulletin', Modern Railways 31:406-7.

5.   C.E. Banister (1981) 'Transport in Romania: a British perspective', Transport Review 1:251-70; D. Turnock

(1978) 'Modernisation on Romanian railways', Modern Railways 35: 416-19; D. Turnock (1979) 'The Romanian railway debate: a theme in political geography', Journal of Transport History 5: 105-21; D. Turnock (1987) 'The development of the Romanian railway network since the First World War', Leicester University Geography Department Occasional Paper 16. On the narrow gauge see D. Turnock (1976) 'Narrow gauge in Romania', European Railways 27(4): 7-10; D. Turnock (1986) 'Industrial railways in Eastern Europe: the case of the forest railways in Romania', in W. Ritchie et al. (eds) Essays for Professor R.E.H. Mellor, Aberdeen: University of Aberdeen, 79-87.

6.   C. Garratt (1984) 'Yugoslavia: a stronghold of steam', Steam Railway 45:36-40; F.B. Singleton and J. Wilson (1977) 'The Belgrade-Bar railway', Geography 62:121-5; O. Wilson (1971) 'The Belgrade-Bar railway: an essay in economic and political geography', in Hoffman (ed.) 365-93.

7.   D.R. Hall (1975) 'Some developmental aspects of Albania's fifth Five Year Plan 1971-1975', Geography 60: 129-32; D.R. Hall (1984) 'Albania's growing railway network', Geography 69: 263-5; D.R. Hall (1986) 'Problems and possibilities of an Albanian-Yugoslav rail link', in Ambler et al. (eds) 206-20; D.R. Hall (1987) 'Albania's transport co-operation with her neighbours', in J.F. Tismer et al. (eds) 379-99; C. Portway (1972) 'Albanian journey', Railway World 33: 451-3; C. Portway (1972) 'Forbidden railways of Albania', Railway Magazine 118: 12-15.

8.   Z. Berman and C.G. Alvstam (1987) 'Investment policy in the Polish transport sector', in Tismer et al. (eds) 328-78; A.H. Dawson (1987) 'Transport and the pattern of settlement in Poland', in Tismer et al. (eds) 306-27; R.H. Funck and J.S. Kowalski (1987) 'Impact of transportation bottlenecks: the Polish case', in Tismer et al. (eds) 292-305; L. Webster and C. Garratt (1984) 'The other Poland', in M. Chamberlin et al. (eds) Great Little Railways, London: BBC Publications, 87-114; M. Zajfryd (1974) 'Poland's big investment will boost freight capacity', Railway Gazette International 130:213-15.

9.   R. Helle (1977) 'Future of rail container transport in Eurasia', GeoJournal 1(3): 55-60; R. North (1987) 'Current developments in transport and traffic between the Soviet Union and Eastern Europe', in Tismer et al. (eds) 270-91.

10.   W. Barczuk (1975) 'Elements of the theory of hinterland and foreland relating to research on Polish ports', Geographia Polonica 31:41-52; B. Kortus (1980) 'Production

and spatial links of Poland's industry with foreign countries' Geographia Polonica 43: 223-30; Z. Taylor (1984) 'Seaport development and the role of the state: the case of Poland', in B.S. Hoyle and D. Hilling (eds) Seaport Systems and Spatial Change, Chichester: Wiley, 217-38; J. Zaleski (1978) 'Poland's maritime economy in the post-World War Two period', GeoJournal 2: 179-90.

11.   R. Kolaczkowski (1974) 'Interport: a new solution to the long tug of war over Szczecin', Radio Free Europe Research Paper Poland 9.

12.   A. Ghenovici (1977) 'Development of Romanian sea transport', Revue roumaine: geographie 21: 15-28.

13.   W. Hinsch (1977) 'The river Elbe international', GeoJournal 1: 45-8.

14.   G. Weigend (1975) 'The Danube river: an emerging regional bond', Geoforum 6: 151-61.

15.   A. Ghenovici (1972) 'Changes of navigation in the Danube defile', Revue roumaine: geographie 16: 203-13; D.R. Hall (1972) 'The Iron Gates scheme and its significance', Geography 57: 51-5; A. Herbst and C. Herbst (1971) 'Landscape changes following the construction of the hydro-energetical complex of the Iron Gates', Geoforum 6: 57-62.

16.   G. Korompai (1971) 'Changes in the structure and direction of expanding commodity transport on the Danube', Geoforum 6: 63-74.

17.   T. Sharman (1983) 'Canal on the Danube delta', Geographical Magazine 55: 317-21; D. Turnock (1986) 'The Danube-Black Sea Canal and its impact on southern Romania', GeoJournal 12: 65-79; D. Turnock (1987) 'The Danube-Black Sea Canal and its impact on southern Romania', in Tismer et al. (eds) 400-28.

18.   E.C. Vespremeanu (1980) 'Evaluation of the present state of the Danube delta land region', Revue roumaine: geographie 24: 47-52.

19.   G.A. Giannopoulos (1984) 'Land transport in Southeastern Europe: existing situation and prospects for improvement', Transport Reviews 4: 1-26; Z. Jelinovic (1981) 'The Yugoslav transport system', Transport Reviews 1: 1-23.

20.   K. Kadas (1980) 'Transport policy and regional development: the Hungarian case', Transport Policy and Decision Making 1: 83-91; L. Urban (1981) 'A survey of Hungarian transport policy', Transport Reviews 1: 305-21.

21.   S. Somogyi (1971) 'Natural endowments of the

Great Hungarian Plain', in Sarfalvi (ed.) 43-4; S. Samogyi (1979) 'Regulated rivers in Hungary', Geographia Polonica 41: 39-53; T. Sharman (1984) 'Taming the Tisza river', Geographical Magazine 56: 21-3.

22.   G. Korompai (1977) 'The effects of the Europa Canal Rhine-Main-Danube', GeoJournal 1(2): 33-44; M. Milkowski (1977) 'Waterways and inland navigation in the light of international law and the CSCE final act', Polish Western Affairs 18: 125-43; H.S. Siedenfus (1987) 'From the Rhine-Main-Danube Canal to the Main-Rhine connection', in Tismer et al. (eds) 429-48; D. Turnock (1979) 'The Europa Canal and its Central European context', Bulletin Society of University Cartographers 13(2): 32-6.

23.   M. Starczewska (1971) Potentialities for Polish-Czechoslovak Co-operation in Transportation, Columbia University: Ph.D. thesis.

Chapter Four

ELECTRIFICATION AND THE PROSPECTS FOR NUCLEAR POWER

Increases in electricity production have been dramatic throughout Eastern Europe and the construction of grid systems, interconnected under Comecon auspices, has made for much wider locational choice for manufacturing industry since power can be made available in all parts of each country at uniform cost to the consumer (Figure 4.1; Table 4.1). (1) The electrification of individual towns and villages shows a clear trend towards greater regional equality. Prior to 1945 there were 443 settlements with electricity in Romania, 57.5 per cent of them in the former Hungarian territories, although their share of the population was only 36.3 per cent. This did not represent any deliberate discrimination but simply the fact that the population in these areas was more highly urbanized with relatively high incomes, coupled with the availability of power through a network of power stations and transmission lines. After 1945 the development of the grid system enabled more rapid progress to be made in other areas. The former Hungarian territories accounted for 42.5 per cent of settlements provided with electricity between 1945 and 1955 but only 38.4 per cent of those connected between 1955 and 1965: only slightly in excess of the population share. By contrast the north-east and south-west, with only 12.2 per cent of connections up to 1945, boosted their combined share to 20.5 and 25.2 over the following two decades.

The electricity supply industry has generated linkages with electrical engineering (supplying the generators, turbines, and transformers). Thus the Vassil Kolarov plant in Sofia produces electric motors and generators while the transformer plant in Sofia produces circuit breakers and

Figure 4.1 Power stations and electricity transmission

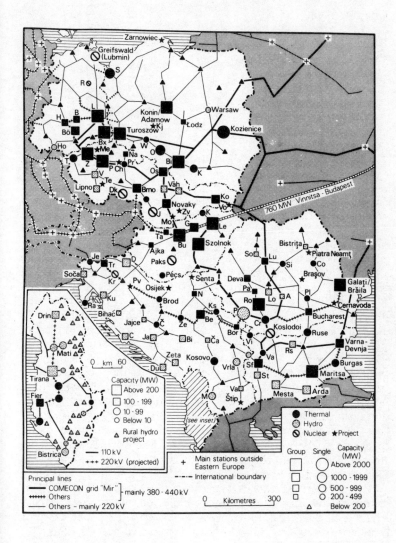

Source: National atlases.

Note: For key to power stations see Note 2 on page 146.

121

Table 4.1 Electricity capacity and production 1960-80

Country	1960 (a) A	B	C	D	E	1970 A	B	C	D	E	1980 A	E	D	C
Bulgaria	4.66	+0.03	0.9	0.5		19.51	+0.10	4.1	0.8		34.83	8.2	1.9	0.9
Czechoslovakia	24.46	−0.26	5.7	0.9		45.16	+3.40	10.1	1.4		72.73	15.6	2.1	0.9
GDR	40.30	−0.38	7.8	0.0		67.65	+0.40	12.1	0.7		98.81	19.8	1.5	1.8
Hungary	7.62		1.4	*		14.54	+3.39	2.5	*		23.88	4.8	*	
Poland	29.31	+0.30	6.3	0.3		64.53	+0.05	13.7	0.7		121.88	24.7	1.3	
Romania	7.65	−0.03	1.8	0.2		35.09	−2.33	7.3	1.2		67.49	16.1	3.5	
Yugoslavia	8.93	−0.09	2.4	1.4		26.02	+0.21	7.0	3.6		59.34	14.0	6.3	
Eastern Europe	122.93	−0.43	26.3	3.1		272.50	+5.22	56.8	8.5	*	478.96	103.4	16.6	3.6
USSR	369.27	−0.23	93.0	20.8	0.1	740.40	−5.35	166.2	31.4	2.0	1,295.00	266.8	52.5	14.0
Western Europe (b)	381.45	+3.67	161.1	34.8	1.2	743.93	+11.85	274.0	90.8	8.2	1,082.13	431.4	120.5	43.0

Source: Statistical yearbooks.

Notes:
(a) Figures for the USSR and Western Europe (except A for Western Europe) are for 1962-3
(b) Covers FRG, France, Italy, and UK
A   Production (Billion kWh)
B   Net trade (billion kWh) (+ indicates net import; − net export). Figures not available for 1980.
C   Total capacity (thousand MW)
D   Hydro capacity (thousand MW)
E   Nuclear capacity (thousand MW)
*   Less than 0.1

voltage regulators. However there is considerable specialization and a large power station in Bulgaria will have to be equipped largely by foreign suppliers, mainly from the Soviet Union, Czechoslovakia, the GDR and Hungary. Power equipment supplied by Czechoslovakia involves turbines of 220MW for nuclear power stations (with 1,000MW turbines in development) by Skoda of Plzen and water turbines by CKD Blansko: the latter include the conventional Kaplan turbines (first built at Blansko by Victor Kaplan in 1913) and multi-stage reversing turbines for pumped storage plants. Meanwhile diesel generators are supplied by CKD Hradec Kralove.

Given the post-war advances in technology and the expansion of electrical engineering in Eastern Europe power station capacities have increased and distribution now involves national and international grids with voltages increasing from 110kV to 220, 400, and 750 during the post-war period (Table 4.2). Power stations must be carefully sited with land stability and cooling water in mind and located with regard to the principal fuel resources. (3) Hard coal and lignite are both very important, especially in the northern countries, particularly with the rising prices of hydrocarbon fuels which must be imported in very large quantities (mainly from the Soviet Union) in view of the limited endowment in Eastern Europe (Table 4.3). Alternatives can be found in hydro-electricity, a very significant option in the Balkan countries where scope for integration with flood control, navigation, and water supply programmes provides added attractions, and also in nuclear power. All the East European countries have now built large power stations based on lignite fields. For example capacities exceeding 1,000MW have been installed at Kosovo in Yugoslavia and Turoszow in Poland; and they exceed 2,000MW in the Belchatow/Konin area of Poland, Rovinari/Craiova in Romania, and Nikola Tesla in Yugoslavia. The greatest concentrations are in Bulgaria, where the eventual capacity of the four power stations at Maritsa Iztok will be 8,400MW, and in the GDR where the capacity of power stations on the Spree in the Cottbus area including Boxberg, Lubbenau, and Vetschau is of a similar order (Boxberg alone has 3,580MW installed).

These power stations and the mining and manufacturing complexes which are located close to them have made considerable modifications to the settlement pattern. At Kazincbarcika in Hungary the lignite quarries, power

Table 4.2 Electrification in Yugoslavia 1945-85

Year	Criterion											
	A	B	C	D	E	F	G	H	J	K	L	M
1945	0.5	0.2	0.3	na	na	na	0.3			0.8	96.3	
1950	0.7	0.3	0.4	0.1	*	*	1.2			2.4	113.2	
1955	1.2	0.6	0.6	0.5	0.3	0.2	2.9			4.3	114.7	
1960	2.5	1.4	1.1	1.2	0.7	0.5	5.3	0.5		8.9	118.8	-93
1965	3.7	2.2	1.5	1.2	0.8	0.4	6.6	2.2		15.5	98.8	+394
1970	6.9	3.6	3.2	3.1	1.5	1.7	8.3	3.7	0.2	26.0	107.0	+209
1975	9.0	4.8	4.2	2.2	1.2	1.0	10.7	4.6	0.6	40.0	90.8	+1,093
1980	14.1	6.1	8.0	5.0	1.3	3.7	13.0	5.0	3.4	59.7	109.9	-441
1985	19.2	7.7	11.5	5.1	1.6	3.5	14.3	5.5	3.7	77.5	61.9	na

Source: Yugoslav statistical yearbooks.

Notes:

A Total power station capacity (thousand MW)
B Total hydro power station capacity (thousand MW)
C Total thermal power station capacity (thousand MW)
D As A but increase over previous five years
E As B but increase over previous five years
F As C but increase over previous five years
G Transmission lines (thousand km) 110 kV
H Transmission lines (thousand km) 220kV
J Transmission lines (thousand km) 380kV
K Electricity output (billion kWh)
L Hydro-electricity: share of output divided by share of capacity
M Net trade in electricity (+ indicates import)
* Less than 0.1

Table 4.3 Energy consumption and composition 1960-90

Consumption per capita GJ	Eastern Europe (a)	Northern group (b)	Balkan group(b)	USSR (b)	Western Europe (b)
1960	76.2	+28.3	-44.8	+6.7	+17.5
1970	116.6	+26.4	-39.9	+13.9	+23.3
1978	145.7	+31.1	-45.8	+16.3	+8.1
1990	165.6	+17.9	-26.1	+85.8	+21.3
Energy Composition (per cent)					
1960 Coal	84.2	+8.1	-36.4	-21.1	-25.5
Oil	8.6	-3.2	+14.2	+17.4	+20.7
Gas	5.8	-4.4	+19.9	+4.1	-4.1
Nuclear					+0.1
Hydro	1.4	-0.5	+2.3	-0.4	+8.8
1974 Coal	59.6	+11.4	-27.7	-23.8	-37.1
Oil	24.4	-4.7	+11.4	+11.4	+32.6
Gas	13.1	-5.9	+14.3	+12.9	-1.1
Nuclear	0.4	+0.1	-0.2	-0.3	+1.3
Hydro	2.5	-0.9	+2.2	-0.2	+4.3

1982	Coal	56.3	+11.2	-24.8	-29.5	-33.3
	Oil	22.4	-3.1	+7.1	+16.1	+26.5
	Gas	16.9	-6.1	+13.5	+13.1	-1.8
	Nuclear	1.2	-0.1	+0.1	+0.3	+4.8
	Hydro	3.3	-1.9	+4.1		+3.8
1990	Coal	53.9	+11.8	-23.5	-27.7	-28.2
	Oil	22.9	-4.6	+9.1	+7.4	+17.0
	Gas	13.4	-3.2	+6.4	+18.1	+1.5
	Nuclear	5.4	-1.0	+2.0	+0.5	+5.6
	Hydro	4.4	-3.0	+6.0	+1.7	+4.1

Sources: UN/ECE (1983) An Efficient Energy Future: Prospects for Europe and North America, London: Butterworth; G.W. Hoffman and L. Dienes (1985) The European Energy Challenge: East and West, Durham, NC: Duke University Press.

Notes: a Actual values
        b Deviations from East Europe profile

station, and chemical complex (Borsod) have given rise to a new town of nearly 40,000 population where previously there were three villages. The same combination has generated an even larger complex at Dimitrovgrad in Bulgaria although much of the power goes to other parts of the country, using the 400 kV transmission ring which was completed in the late 1970s (serving Burgas, Plovdiv, Sofia, and Varna). By contrast early power station developments in Bulgaria rested on the supply of hard coal from the Soviet Union. Substantial quantities still move from Zhdanov to Varna and, in the case of Ruse power station on the Danube, from the Soviet river port of Reni. Oil and gas have also been used as power station fuels but high cost now prohibits this from continuing on a large scale. Despite the fact that gas deliveries from the Soviet Union seem assured over the long term, the cost of gas makes it an unattractive fuel for electricity generation and much of the imported gas is likely to find its way into chemical plant. Even in Romania where gas has been used extensively for power production there is now a swing towards greater use of low-grade coal and expansion of hydro and nuclear capacity. (4)

HYDRO POWER

Hydro-electricity was given great importance in the Balkan countries in the early post-war years because there was considerable potential and a shortage of high-grade fuels (except in Romania). Large-capacity thermal stations subsequently proved most attractive but in the context of present fuel costs the hydro potential is again attracting much attention. Between 1945 and 1985 Yugoslavia's hydro-electric power station capacity has risen 36.5 times from 211 to 7,705MW while the increase in thermal (including nuclear) capacity has been 34.5 times from 313 to 10,787MW. But it is noticeable that while hydro capacity increased rapidly in the early post-war period the growth of thermal capacity has been evident in recent years. However rising fuel costs are leading to a closer consideration of further hydro schemes, for only some 40 per cent of the potential has been exploited so far and there are still many opportunities on the Drina, Morava, Sava, and Zeta rivers. Further developments are also possible in the other Balkan countries.

Pumped storage schemes offer advantages in smoothing

out sharp daily variations in demand. Such schemes date back to the inter-war years in Germany: projects were implemented on the Elbe (Dresden) and Saale (with some incidental benefits to navigation) and there is now a plan for the Erzgebirge which can take advantage of surplus thermal power available at off-peak periods. Yugoslavia has pumped storage hydro schemes at Bajina Basta (592MW) and Capljina (210MW) while Czechoslovakia's pumped storage hydro-electric schemes are at Cierny Vah (660MW), Dalesice (410MW), Dlouhe Strane (600MW), and Liptovska Mara (198MW). Dalesice is closely linked with the Dukovany nuclear power station and will use off-peak nuclear power to pump water to the upper reservoir. However the project will also assure adequate cooling water from the Jihlava river to the nuclear power station at all times of year and there will be additional benefits in irrigation and flood control. An element of pumped storage is involved in Romania's Lotru scheme since off-peak power is used to lift water from reservoirs outside the main catchment. However in contrast to large complex schemes small conventional hydro-electric stations are once again being looked on with favour. A construction programme has been launched in Yugoslavia where small power stations with capacities of about 1 megawatt will be useful for supplying small rural industries and similar plants are being restored in Czechoslovakia to reduce pressure on the national grid system.

Hydro-electricity offers the possibility of bringing output into line with consumption in regions which lack the fuels to support large thermal power stations. The Yugoslav republics with the highest per capita power consumption in 1980 (Slovenia with 1.86 times the national average and Montenegro with 1.67) are both able to produce most of their own power (with per capita outputs respectively 1.59 times and 1.44 times the national average) thanks to the hydro potential. The surpluses in Bosnia and Hercegovina (where output is 1.02 times the national average but consumption 0.83 times) and Serbia (1.53 and 0.93) arise in part through hydro power, while Macedonia's deficit (0.38 and 0.87) is moderated by the hydro potential. Regions with little hydro potential are heavily in deficit: Croatia, with values of 0.75 for output and 1.05 for consumption and Vojvodina with an even greater disparity: 9 and 120. Not that self-sufficiency is necessarily desirable as a matter of course. Inter-regional or international trade is necessary when the exploitation of major sites will give rise to temporary surpluses while seasonal factors also make exchanges appropriate. Yugoslavia's net trade in electricity has increased from an export of 93 million kWh in 1960 to an export of 441 in 1980 but has decreased from 1.0 to 0.7 per cent as a proportion of the total amount of power generated. However the total turnover has increased both absolutely (from 310 to 5,954 million kWh and proportionately 3.5 to 10.0 per cent of total power output) through a major import from Albania and exports to Austria, Bulgaria, Greece, Italy, and Romania.

Hydro-electric schemes can of course be very disruptive. Many villages were affected by the Bicaz scheme on the Bistrita river in Romania and similar consequences have followed from the chain of sixteen dams built on the Vah in the Slovakia since 1950. The system culminates in the Liptovska Mara serving a 210MW power station. Thirteen villages were flooded and people had to be rehoused in nearby towns, especially Liptovsky Mikulas (where the former church has been retained as a museum) and Ruzomberok. However there are local benefits arising through flood control, irrigation, and a tourist industry has arisen because of the scope for water-based recreation. A similar scenario can be noted on other rivers in Czechoslovakia. The harnessing of the Vltava has resulted in dam construction extending upstream from the first dams of Vrane and Stechovice to Slapy, Kamyk, Orlik, and finally Lipno. The dam, built on slopes of Vihorlat in East Slovakia, forms part of an extensive drainage and irrigation scheme which eliminates the danger of spring flooding on the low ground. There is a holiday centre on the banks of the reservoir. In Bulgaria electrification has integrated with irrigation in the Maritsa valley thanks to the construction of some major mountain storages (behind the Batak, Pjasatchnik, and Topolnitsa barrages) and some 300 small reservoirs scattered across the Sredna Gora and Rhodope. During the April-September irrigation season a range of crops is grown: cereals and rice, sunflowers, lucerne, tobacco, and various orchard and vineyard crops. Pollution is a growing problem however and the Maritsa is now polluted throughout its length, whereas until 1960 there were no difficulties upstream of Pazardjik. (5) Bulgaria is now working on a major hydro-electric complex, Belmeken-Sestrino, involving four power stations with an eventual total capacity of 2,200MW. The system will collect water over an area of some 4,000 square kilometres.

NUCLEAR POWER

During the 1970s the rising world price of oil and the pressure from the Soviet Union for capital investment from Eastern Europe to open up oil and gas fields in remote parts of the USSR increased interest in nuclear power, although the origins lie further back in the 1960s, when Eastern Europe as a whole found it increasingly difficult to find enough domestic energy resources to sustain an ambitious programme of industrialization. Nuclear power has been seen (albeit controversially) as the best hope for increased electricity output in a way which is both economically and ecologically acceptable. Thus the 1990s are seen as an appropriate time for a major advance (Table 4.4). However technical problems may lead to further delays while the effects of both falling world oil prices and the Chernobyl disaster in the Soviet Union have yet to be felt. (6)

The development of nuclear power may be traced back to uranium mining at Horni Slavkov (Jachymov) in Czechoslovakia immediately after the Second World War. The place became notorious for the inhuman conditions in which German prisoners and the regime's 'class enemies' were forced to work. Deposits were depleted by the 1960s and operations ceased in 1967. Production now comes from Pribram, where the uranium enterprise was founded in 1949, and deposits have been reported at Ceske Budejovice, Ceska Lipa, Tachov, Zbraslav, and Zdar. There has also been activity around Aue in the GDR and it is believed that small quantities are mined in Bulgaria and Hungary. There may be some production in Romania, for the Russians were active in the Vascau area of Bihor in the 1950s and the two new towns of Dr Petru Groza and Nucet are a legacy of this initiative. But the Romano-Soviet agreement on uranium deliveries was cancelled at the beginning of the 1960s. Meanwhile Czechoslovakia was advancing further since the Soviet Union provided essential assistance in building the Czechoslovak Institute of Nuclear Research at Rez in 1955 and trained hundreds of Czechoslovak workers for the nuclear industry. The research reactor, with capacity raised from 2MW (1957) to 4MW, and later 10MW, produced the first isotopes in 1958.

Being well endowed with natural uranium and wishing to be independent of an outside supply of enriched uranium Czechoslovakia favoured a commercial reactor using natural uranium fuel, carbon dioxide coolant, and heavy water

Table 4.4 Energy production and consumption 1973-2000

Production (A) and trade (B) in million tonnes of standard fuel

Country	Year	Solid A	Solid B	Liquid A	Liquid B	Gas A	Gas B	Electricity AA	Electricity AB	Electricity B
Bulgaria	1973	230	+159	8	+485	8			25	+12
	1980	254	+192	9	+562	4	+195	70	37	+14
	1990	380	+192	90	+818	210	+334	165	28	+25
Czechoslovakia	1973	1,808	-9	7	+599	31	+77	2	22	+16
	1980	1,919	-49	4	+789	19	+249	44	42	+7
	1990	1,924	-23	4	+844	47	+405	301	52	+22
	2000	1,902	+26	4	+876	54	+451	744	71	+26
GDR	1973	2,264	+255	3	+555	92	+27	5	17	
	1978	2,249	+227	4	+733	117	+126	71	4	+5
	1990	2,726	+84		+649	96	+201	268		+13
Hungary	1973	352	+98	83	+250	173	+8		1	+17
	1980	305	+85	102	+338	225	+120		2	+27
	1990	365	+70	80	+340	230	+135	120	20	+35
	2000	485	+55	90	+345	230	+125	310	20	+35
Poland	1973	4,093	-1,077	16	+514	178	+50		8	
	1980	4,813	-831	15	+710	147	+200		15	
	1990	5,098	-732	15	+760	147	+200		15	
	2000	5,860	-679	15	+1,000	147	+350	45	15	

Romania	1973	295	+41	578	-25	1,018	-8		79	
	1980	530	+112	540	+269	1,081	+35		126	-7
	1990	1,000	+168	500	+421	870	+159	191	270	
Eastern Europe[a]	1973	9,042	-533	695	+2,378	1,500	+154	7	152	+50
	1980	10,120	-310	672	+3,301	1,582	+996	236	224	+55
	1990	12,021	-389	689	+3,972	1,600	+1,484	1,045	385	+95
	2000	13,663	-335	600	+5,346	1,550	+2,300	2,597	586	+140
USSR	1973	13,821	-590	18,087	-4,343	8,326	+200	147	1,256	-35
	1980	14,239	-251	25,255	-6,869	15,277	-1,622	590	1,770	-83
	1990	19,908	-1,475	27,016	-5,692	26,396	-4,217	4,129	2,359	-147
	2000	21,825	-3,244	25,659	-3,244	25,686	-5,014	8,828	2,507	-295
Western Europe	1973	8,353	+1,036	511	+22,886	4,780	+172	724	995	+33
	1980	7,767	+1,530	3,837	+17,023	5,378	+1,939	1,520	1,265	+85
	1990	7,594	+3,407	4,406	+18,377	5,255	+4,542	6,349	1,364	+13
	2000	8,232	+5,040	4,400	+18,500	3,443	+5,600	10,031	1,550	

Source: UN/ECE (1983) An Efficient Energy Future: Prospects for Europe and North America, London: Butterworths.

Notes: AA Nuclear
AB Hydro
a Excluding Albania and Yugoslavia

Table 4.5 Nuclear power stations

Country/location	Reactor type and capacity			Progress			
				A	B	C	D
Bulgaria							
Kozlodoui	VVER	4 x	440 MW	1966	1969	1974	1982
	VVER	2 x	1,000 MW	na	1981		
Czechoslovakia							
Jaslovske	GCHW	9 x	50 MW	1956	1958	1960	1972(a)
Bohunice	VVER	2 x	440 MW	1970	1973	1978	1981(b)
Dukovany	VVER	2 x	440 MW	1970	1977	1983	1986
	VVER	2 x	440 MW	1970	1974	1980	1986
	VVER	2 x	440 MW	1970	1979	1984	1987
Mochovce	VVER	4 x	440 MW	1978	1980	1990	
Temelin	VVER	4 x	1,000 MW	1981	1984	1995	
Melnik	VVER	4 x	440 MW				
Chvaletice	VVER	4 x	1,000 MW			2000	
Vojany	VVER	2 x	1,000 MW			2000	
Zvolen	VVER	2 x	1,000 MW			2010	
GDR							
Rheinsberg	PWR	1 x	80 MW	1960	1961	na	1966
Greifswald	VVER	4 x	440 MW	1967	1967	na	1979

				A	B	C	D
Hungary							
Paks	VVER	4 x	440 MW	1966	1972	1980	1988
Poland							
Zarnowieckie	VVER	2 x	440 MW	1975	1981	1988	
Romania (c)							
Cernovoda	Candu	1 x	700 MW	1978	1980	1985	
Yugoslavia (d)							
Krsko	Westinghouse	1 x	664 MW	1973	1974	1978	1981

Source: Radio Free Europe Statistical Reports.

Notes:
a   Only two reactors were completed before the abandonment of the project in favour of VVER in 1970 and it is understood that both were scrapped in 1979.

b   It was originally planned to have two 440 MW reactors at Jaslovsky Bohunice and Dukovany but capacity was doubled and the grouping of four reactors has now become standard practice at other locations.

c   Other locations announced: Brasov and Piatra Neamt.

d   Other locations announced: Osijek, Prevlaka, Senta, Stip.

A   Order placed
B   Construction started
C   Planned completion date
D   Actual completion date

GCHWR   Gas-cooled heavy water reactor
PWR   Pressurized light water reactor
VVER   Russian acronym for PWR

moderation. Plutonium yielded by this process could then be used for second-generation (fast breeder) reactors. So a further agreement with the Soviet Union provided for the construction of the first commercial nuclear power plant at Jaslovske Bohunice (150MW) built between 1958 and 1972 involving a gas-cooled heavy water reactor (GCHWR) designed by the Czechs. But it was only in 1971 that the USSR moved swiftly to complete installation, by which time the project was overtaken technologically by the Soviet 440MW light water pressure reactor (PWR) and since 1969 Czechoslovakia's programme has been geared to this new technology which requires less investment per unit of capacity. Thus the GCHWR programme has been curtailed in favour of the PWR for which a Soviet supply of slightly enriched uranium is needed. Both Poland and Romania had by this time given up their own plans for GCHWRs while the other Comecon countries accepted Soviet offers of PWRs from the start (see Table 4.5). The Soviet attitude is not altogether unreasonable. There is much to be said for avoiding a proliferation of reactor types and assistance to Eastern Europe for HWR development would have been difficult when the Russians themselves had installed no such plant of their own. The situation was not helped by the NATO embargo on strategic goods which both Czechoslovakia and Romania found to be virtually impenetrable until the very end of the 1960s.

National profiles

Czechoslovakia's new nuclear programme began with an agreement with the Soviet Union in 1970 for four of the 440MW PWRs. It was originally planned that reactors would be located in pairs at Jaslovske Bohunice and Dukovany but groups of four were subsequently preferred. A cluster of four 440MW reactors has now been installed at Jaslovske Bohunice beginning in 1978. Similar installations are being made at Dukovany in Moravia (and Mochovce near Levice in Slovakia) where work that started in 1974 has been slowed down until installation of the 1,760MW complexes could proceed. (7) But the basic type 440MW reactor is now superseded by the fast breeder reactor, which has been operating in the USSR since 1973, leading to the powerful 1,000MW reactor which went on stream in 1980 (at Novovoronezh). This should effect the transition to a larger-scale nuclear power industry which is self-sufficient in fuel; for in 1980 the Comecon countries signed an agreement on the development of powerful fast breeder reactors with capacities ranging from 800 and 1,600MW. However, the immediate goal is the series production of the Novovoronezh hot electron reactor. In 1980 it was agreed that the 1,000MW reactors would be installed at Temelin and Melnik in Bohemia with four units at each location, making these power stations the largest in Czechoslovakia by 1990. Hot water from nuclear power stations is being used for domestic heating. A pipeline runs from Jaslovske Bohunice to Trnava and another will run from Dukovany to Brno. This will effect substantial savings of gas and oil. Meanwhile uranium is still being mined at various places in Czechoslovakia (chiefly Hamr near Ceska Lipa but also Nove Mesto na Morave and Pribram) and total annual production exceeds 2,000 tonnes. Work is done under Soviet supervision and all production not required for the domestic nuclear programme is exported to the USSR.

Plans for nuclear power in Bulgaria and Hungary date back to 1966 and involve the use of the Soviet 440MW reactor. However there have been some contrasts between the two countries. In the case of Bulgaria two units were installed at Kozlodoui between 1969 and 1975. This was the first nuclear power station in the Balkans and capacity was doubled by 1982. Then as the Russian 1,000MW reactor came on the scene in the mid-1970s Bulgaria anticipated having an additional 2,000MW of nuclear capacity available by 1980 and 8,000MW by 1990, by which time half the electricity produced in the country would be from nuclear sources. Progress has not been quite so rapid but in 1981 preparations were being made for the erection of the first 1,000MW reactor at Kozlodoui. By contrast the Hungarian project for Paks, on the Danube south of Budapest, has not run smoothly. (8) After an initial period of optimism (1966-9) the project was virtually ignored by the media due to a decision to postpone construction. The scheme was revived in 1972 and construction resumed with all speed in 1974 with ultimate capacity doubled. The station was originally due for completion in 1975 but the target was subsequently revised to 1980 (in 1970) and the first of four reactors was commissioned in that year, with the three others following on through to 1988 when Paks should be responsible for a third of all the electricity generated in Hungary. Although a more adequate supply of hydrocarbon fuels was given as the reason for postponement in the early 1970s it seems that the real reason lay with concern about the safety of the plant: the Hungarians evidently laid down stricter standards than the Russians. The Hungarians now plan to install two of the 1,000MW reactors at Paks in the late 1980s.

The GDR and Poland, with large coal reserves, have not been particularly enthusiastic about nuclear power. The GDR's first nuclear power station opened at Rheinsberg in 1966, six years late. This was an 80MW PWR and it has been followed by a development at Greifswald involving four 440MW PWRs completed between 1975 and 1979. Poland's nuclear power project at Zarnowieckie involves four 440MW PWR-VVER reactors ordered in 1975. Construction began in 1981 and commissioning is planned for 1988. A cluster of four 1,000MW nuclear reactors is planned for Kujawy. Poland has however joined with her Comecon partners in financing 4,000MW nuclear power stations in the Soviet Union. The first was started at Khmelnitskiy in the Ukraine in 1979 and the second was mooted for Konstantinovo in 1981; 750kV transmission lines will take power into Poland and also across Romania into Bulgaria.

Yugoslavia's first nuclear power station (664MW) opened at Krsko on the borders of Croatia and Slovenia in 1981. These two republics have been seriously affected by periodic power shortages. Assistance has been forthcoming from Westinghouse who have also participated in the construction of the Slovenian uranium mine of Zirovski Vrh at Skofja Loka, and in the construction of gas-burning power stations at Jertovec in Croatia (92MW) and Trbovlje in Slovenia (62MW). Yugoslavia imported uranium for the Krsko reactor until 1982 when the Zirovski Vrh uranium mine came on stream. Since then Yugoslavian uranium has been sent to the USA for enrichment. Romania's targets have not yet been achieved. There is still no output from nuclear power stations in spite of the emphasis placed on research and development during the Ceausescu epoch. Electrification plans drawn up in the 1960s and 1970s envisaged nuclear power capacities of 1,000 and 2,400MW in 1970 and 1980 respectively. But it was only in 1970 that agreement was reached with the Soviet Union for the installation of a 440MW reactor at Cernavoda. Work began in 1972 but proceeded only slowly until operations were apparently abandoned in 1978 (following the cracking of the concrete foundations during the earthquake in 1977). In 1978 however Romania purchased a Candu reactor from Canada at a cost of 400 million dollars. It was to be installed at Cernavoda by 1985 but although work began in 1980 the scheme remains incomplete.

The early 1980s saw Romania attempting to do business with both Canada and the Soviet Union. Plans were discussed for a package involving as many as sixteen Candu reactors with a cluster of four or five at Cernavoda and another group in Transylvania. However momentum was lost in 1980 when financial difficulties held up the signing of a deal for a second reactor. Both Canada and the USA subsequently suspended credits for the Romanian nuclear power programme in 1982. Meanwhile plans were advanced to install Soviet reactors in Moldavia with a view to achieving a total nuclear capacity of 4,500MW in 1990 and 9,600MW in 1995. With a parallel development of hydro power it was envisaged that the thermal capacity would be reduced from 82 per cent of the total in 1980 to only 58 in 1990. But this target seems far too ambitious and the planned electricity output for 1990 has come down from 110.0 billion kWh envisaged in 1982 to just 101.7 (which is still high in relation to the figure of 71.8 achieved in 1985). However the nuclear project proceeds slowly. In 1985 it was reported that sites for the Moldavian and Transylvanian power stations had been established near Piatra Neamt and Brasov respectively while nuclear power featured in the programme of the Romanian president's visit to Canada during the same year. In 1986 the manufacture of major items of equipment for the Cernavoda power station was mentioned (specifically a 440MW transformer by the Electroputere works at Craiova). The balance between Soviet and western dealings is of course typical of Romanian policy. A deal restricted to Canada must have appeared financially unsupportable to say nothing of the problem of uranium supply. The Candu reactor's dependence on natural uranium avoided the problem of enrichment facilities, and also opened up the prospect of eventual construction of such reactors in Romania for export to the Third World. But there appear to have been difficulties with the joint company set up by Romania and Zambia to develop mines at Maokambo to yield coal and nickel as well as uranium. Hence the co-operation agreement with the Soviet Union reactivated in 1982. The Russians will supply enriched uranium and also handle the reprocessing of spent fuel.

Engineering

The Comecon countries have pooled their efforts in research and design since 1979. Activity in building nuclear equipment involves about fifty enterprises, whose efforts are co-ordinated by Interatomenergo. The Soviet Union makes the greatest contribution through units like the Atommach plant at Volgodon and the turbine plant at Kharkov. Czechoslovakia also plays a key role, since her relatively early involvement in the nuclear power industry, along with the proven competence of firms like Skoda, has given the country a key role in the supply of equipment for nuclear power stations throughout Comecon. A large investment programme has been undertaken at Plzen to enable Skoda to assemble various Soviet components (notably reactors) for 440MW reactor vessels, and subsequently 1,000MW vessels of the Novovoronezh type. Klement Gottwald works at Vitkovice will produce high-pressure vessels and steam generators (in collaboration with the Soviet boiler works in Taganrog); piping and turbines will be provided by Valcovny Trub of Chomutov; pipes and pumps by Sigma of Olomouc; and other work by the Slovak energy engineering enterprise at Tlmace. Czechoslovakia is to supply twenty-one 440MW reactor vessels, ten for domestic power stations and eleven for export, before switching to the 1,000MW. The GDR and Poland are emerging as specialized producers of certain types of nuclear sub-assemblies and instruments including rust-proof heat exchangers while Hungary has produced air-conditioning devices and water-purification installations. Using her principal engineering plants like Vassil Kolarov in Sofia, Bulgaria is developing a number of small parts including air-tight doors; she is also responsible for biological protection systems. Romania has some involvement with transport equipment. Another Comecon organization, Interatominstrument, co-ordinates the production of instruments not only for nuclear power stations but also for use in radiation protection, radium therapy for cancer patients and other medical treatments. The Soviet firm Izotop is prominent along with others from Eastern Europe including Tesla (Czechoslovakia) VEB Robotron (GDR), Gama (Hungary), and Polon (Poland). Nuclear engineering industries could eventually attract orders from the Third World and Czechoslovakia would be in a good position in this respect.

Nuclear safety

Nuclear safety is now being taken very seriously. While it is a matter for satisfaction that Kozlodoui was not affected by the 1977 earthquake it seems that some accidents have occurred in Eastern Europe, particularly in Czechoslovakia where Charter 77 publicized incidents at Joslovske Bohunice in 1976 and 1977 relating to emissions of carbon dioxide and radioactive gas. Such releases seem to have been part of a series of difficulties over the building and operation of the GCHWRs, which have now been scrapped. There was some loss of life at the power station and some international repercussions, especially in Austria, whose border with Czechoslovakia lies close to both the power stations now operating as well as Mochovce and Temelin, where construction work is proceeding. Pollution has also occurred through inadequate supplies of cooling water and through uranium mining operations. All this causes concern at home and abroad. It is therefore important to ponder the significance of major accidents (Harrisburg and Chernobyl) for the future of nuclear power in Eastern Europe. There is no doubt that certain repercussions will be felt since, at the grass roots, there is obvious anxiety about the risks inherent in a nuclear policy. The accident at Harrisburg in the USA in 1979 was widely reported in Eastern Europe but in a rather perfunctory manner, presumably to avoid creating anxiety about nuclear projects under way. Chernobyl in 1986 made more of an impact because people (especially in Poland) were involved in medical precautions.

The media in Hungary and Yugoslavia has stimulated frank discussion about nuclear safety however, and further awareness of the dangers in Yugoslavia arises from the presence of Yugoslav Gastarbeiter in western Europe, where extensive discussions and demonstrations can be observed. Protest has been particularly effective in Yugoslavia, where the opposition from Zadar against the proposal of the Croatian authorities to site a nuclear power station on the island of Vir led to the decision to select another location (Prevlaka on the River Sava) in 1979. The initial location was determined in broad terms by the energy deficit in the Split area which contains the largest complex of industries in Dalmatia. But Zadar was concerned about its tourism which has developed on the strength of stable weather conditions, clean water, and an abundant supply of fish, fruit, and vegetables. It was felt that fear of nuclear radiation could damage the tourist industry and affect the ecological balance of the area. Similar concern was expressed at Sibenik. Interestingly the power of regional interests was again demonstrated in Yugoslavia in 1979 when the authorities in Vojvodina successfully defended their interest in two oil-burning power stations which the Federal Assembly sought to cancel as part of an energy conservation measure.

The Hungarians have also become very safety conscious especially after the Three Mile Island accident in the USA. Although the decision to delay the Paks project for five years (1970-5) and then to restart work earlier than expected (1973) has been attributed to the ready availability (and subsequent rise in price) of hydrocarbon fuels, it may also be related to concern over safety and lack of adequate skilled labour in Hungary. The Paks station has some special safety features including fall-out monitoring stations and mobile laboratories. Waste material is stored in special containers submerged in water pending shipment to the USSR. These safety features have evidently added considerably to the total costs. The Poles have also been much concerned over nuclear safety and for a time the Chernobyl accident disrupted the export of vegetables from eastern to western Europe. However the official view remains one of steadfast support for the nuclear option, although there will obviously be more consideration given to safety measures, including the training of power station workers. The Russians will join with the Czechs in working out a system of reactor control and safety and with the Poles on the question of security.

Of course there is not the same intensity of public debate in eastern Europe as is encountered in the west but there is little doubt that nuclear power is proving extremely costly and the substantial delays in completing construction reflect the difficulties in providing an adequate infrastructure. The long delay in commissioning the first reactors at Jaslovske Bohunice points to a remarkable underestimation of the technical and research problems. Investments have been higher than planned and labour supply has been a serious problem. It is not just a question of training nuclear scientists and power station workers, for assembling and maintaining an adequate construction staff has been difficult. A shortage of carpenters and welders has caused delays at Dukovany and the labour crisis at Paks led to the mobilization of Hungarian soldiers and young people and the import of teams of welders from Poland in a bid to minimize slippage. The largest number of workers at Paks at any one time was 10,000! The Czech programme is particularly ambitious and shortages of building workers, including carpenters and welders, have been reported. Work is going on simultaneously at various sites and, it seems, with a surprising lack of standardization: the pattern of work at one site differs greatly from that at the others. The question of costs has to be seen in the light of other ambitious electrification plans. Since the Paks project was first started the Hungarians have committed resources to the Nagymaros-Gabcikovo hydro-power project on the Danube, in collaboration with Czechoslovakia, and to an expansion of coal-mining to support new power stations at Bicske and Visonta as well as equipment for secondary oil extraction at Nagylengyel. All this to avoid dependence on a single source of electricity. But the investment costs are high and are additional to commitments in the USSR in connection with the Khmelnitskiy and Konstantinovo plants and associated transmission lines.

SOME OTHER SOURCES OF ENERGY

Economical use of energy is now a pressing necessity for eastern Europe but it will take a long time to make major savings. Industry has been geared to extravagance in energy consumption: thus the GDR consumes 7.6 tonnes of coal equivalent energy per capita compared with 6.0 in the FRG. Everywhere a change to less energy-intensive production is needed while staggered rest days have been resorted to in Bulgaria in order to even out demand. Czechoslovakia is working on more efficient power station equipment and on new types of trolley buses. Semi-toroidal picture tubes could reduce the power input of television sets by a third and domestic heating could be reduced by the same margin with thermal insulation. In the meantime the private consumer, who is already hard-pressed, is being called upon to make further sacrifices through power cuts, inferior street lighting, reduced television programmes, and restrictions on the use of private cars. Some progress is being made in the use of surplus heat (from power stations and compressor stations) and renewable energy resources such as solar and wind energy. Warm water from Tisova power station in Czechoslovakia is being used for fish breeding. Pumping and filtering equipment has been installed to supply nearly a hundred ponds for the rearing of trout. Waste heat from gas pipeline compressor stations is being used in greenhouses near Breclav. Nineteen co-operatives pooled resources to use the heat for a joint fruit, vegetable, and wine-growing enterprise based at Velke Bilovice. There are 2.2 ha of greenhouses and 3.2 ha of frames with plastic sheeting producing flowers and vegetables. At the Kourim station waste heat is used to dry fodder. Fish breeding is another possibility.

Solar installations provide some heating and hot water in apartment blocks in Timisoara and elsewhere. A solar installation for heating 2,400 litres of water operates at the Tirgu Lapus machinery station, and is the first of a series of such installations to be completed in the Maramures. A Czechoslovak enterprise based at Mohelnice is producing windmill generators for weekend cottages: an accumulator is charged once the wind speed reaches 3.5 metres per second. And Romania is conducting experiments wih wind-powered generators near Voineasa in the Southern Carpathians. Considerable liquid fuel can be saved by using agricultural by-products (straw and corn/sunflower stalks) and refuse (dried vine tendrils and wood shavings), one-third of which could be used to save oil. Agard and Babolna state farms have used American and Austrian technology to set up their own power plants and the scope for further economies would appear to be substantial: 1 tonne of straw can dry 5 tonnes of maize and save a quarter of a tonne of fuel oil in the process. Applied to the drying of the entire Hungarian maize crop this would mean a saving of 0.4 million tonnes of fuel oil. However although many co-operatives have sufficient raw material they lack the finance to install the equipment and subsidies will be needed if rapid progress is to be made. On this basis a hundred large agricultural enterprises could be saving 0.3 million tonnes of fuel oil by the late 1980s. Meanwhile some progress is being made with biogas installations.

The Hungarians are using geothermal energy from the Great Plain (especially Csongrad). There are now some 600 thermal wells with a total output of 200 million cubic metres per annum. The water was first used for horticultural purposes in the early 1950s and now some 90 per cent of all greenhouses and folia coverings are heated in this way. The warm thermal waters of Szentes are used to rear vegetable plants which can be sent out to the large farms (co-operatives and state farms) in the area as well as household plotlands for planting in unheated tents once the risk of frost is over. Thus early crops of cabbages, cucumbers, peppers, and tomatoes can be grown at relatively low cost. The first produce is ready by the middle of April. Other agricultural uses have emerged in stock-rearing and poultry farms, fish farms, absorption refrigerators, and crop driers. Block of flats in Szeged have been connected to the geothermal network for heating and hot water supply since 1960 and supplies are now made available to factories and hospitals as well. Since the community sector requires the service for only about 200 days per year there is a summer surplus which can again be used in agriculture and the production of mushrooms in the fields, under folia coverings, has now taken off. Further use of geothermal energy is planned and an advanced reinjection technology has been perfected which allows for the recirculation of the water without environmental problems. (9) Thermal waters have also been tapped just across the border in Romania to heat buildings in Jimbolia and Teremia Mare.

Romania is pioneering the exploitation of low-grade bituminous schist (with only one-third of the calorific value of material used in the Soviet republic of Esthonia). Deposits are known in the Banat Mountains, Northern Transylvania, and the Subcarpathian of Moldavia but only in the former case have detailed investigations been carried out: 300 million tonnes can be obtained by quarrying at Doman near Resita while at Anina double that quantity is available by quarrying with a further 3.5 billion tonnes accessible through mining. In the first instance the bituminous schist will be worked in quarries to the east, north, and west of Anina. The northern quarry will impinge on the town itself but because of environmental difficulties in the vicinity of all the quarries (exposure to dust and liability of structural damage during blasting) the entire town will be rebuilt on rising ground well to the north of the quarry operations. The new town (adjacent to the hamlet of Celnic) will accommodate some 20,000 people with an economy based on mining, manufacturing, and tourism. The western quarry has already been developed, having been selected because it involved least disruption of the infrastructure. Some 10 per cent of the material is discarded at the preliminary sorting stage leaving the remainder to be subjected to pyrogenation (to remove gas and oil). The schists, now in semi-coked state, are burnt in the power station along with low-grade fuel oil but natural gas will be substituted when the pipeline is extended from Resita to Anina. Large quantities of ash can then be processed to yield alumina and other material suitable for cement production. However a substantial residual element will have to be dumped along with vast quantities of overburden and the siderite removed at the sorting stage (although in the latter case some economic use may be found and therefore the material is directed to a separate dump). The project is certainly dramatic in landscape terms with a band of quarrying extending along a 2-kilometre belt from Sommerfrische through Ponor to Crivina where the power station is situated (on an elevated site which has been difficult to supply with cooling water). When the eastern quarry has been opened up on the other side of the suburb of Steierdorf (extending from Sigismund to Uteris) and the northern quarry has overwhelmed the town centre the annual output of schist will be 13.8 million tonnes each year and the 400MW power station will run at full capacity.

NOTES

1. A. Borcu (1971) 'The power economy of the Great Hungarian Plain', in Sarfalvi (ed.) 79-87; G. Enyedi (1975) 'Development regions on the Great Hungarian Plain', in Burghardt (ed.) 65-74.

2. Key to power stations: A Arges; B Bitterfeld, Vockerode, Wolfen, and Zschornewitz; Be Belgrade and Pancevo; Bi Bistrica and Basia; Bl Blachownia, Chorzow, Jaworzno, Rybnik etc.; Bo Bohlen, Espenheim, and Thierbach; Bu Budapest, Szazhalombatta etc.; Bx Boxberg, Hirschfelde, and Schwarze Pumpe; C Cetina and Krka; C Catici; Ca Cacak and Kraljevo; Ch Chvaletice; Co Comanesti and Gh. Gheorghiu-Dej; Cr Craiova; D Drava; Dk Dukovany; Du Dubrovnik; G Gyongyos, Gyongyosvisonta, and Lorinci; H Halle, Leuna, and Schkopau; J Jaslovske Bohunice; Ja Jablanica; Je Jesenice; K Krakow; Ka Kazinbarcika and Ozd; K Kosice; Kj Kujawy; Kr Krsko; Ks Kostolac; Ku Kupa, Dobra, and Senj; L Lubbenau, Peitz, and Vetschau; Le Leninvaros and Tiszapalkonya; Lu Ludus; M Mavrovo; Me Melnik; Mo Mochovce; N Novi Sad; Na Nachod and Trutnov; O Opole; Os Ostrava and Detnarovice; P Prague and Dolni Berkovice; Pa Paroseni and Vulcan; Pl

Ploiesti; Pr Pardubice and Opatovice; Pvr Prevlaka; R Rheinsberg; Ra Rasa; Ro Rogojelu and Turceni; Rs Rositsa; S Szczecin (Dolna Odra); Sf Sofia and Pernik; Si Singeorgiu de Padure; So Somes; St Struma; T Tisova; Ta Tatabanya and Oroszlany; Te Temelin; Tr Trbovlje; V Vltava; Va Vardar; Vi Vidin; Vo Vojany; W Wroclaw; Z Zaluzi, Ervenice, Komarany, Pocerady, and Vernerov; Ze Zenica and Lukavac; Zv Zvolen.

3.   K. Dziewonski (1971) 'Research for physical planning in Poland', Geographia Polonica 32: 31-43; C. Mejro (1975) 'Power industry infrastructure in physical planning, Geographia Polonica 32: 53-61.

4.   F.W. Carter (1970) 'Natural gas in Romania', Geography 55: 214-20.

5.   L. Ziapkov (1974) 'Economic activities influencing the fluvial regime of the Pazardjik and Plovdiv plains', in Pecsi and Probald (eds) 157-61.

6.   S. Wasowski (1969) 'The fuel situation in Eastern Europe', Soviet Studies 21: 35-51.

7.   F.W. Carter (1986) 'Nuclear power production in Czechoslovakia', Geography 71: 136-9; F.W. Carter (1988) 'Czechoslovakia: nuclear power in a Socialist society', Environment and Planning 6C: 269-87; R.H. Mathieson (1980) 'Nuclear power in the Soviet bloc', Annals Association of American Geographers 70: 271-9; J. Wilczynski (1974) 'Atomic energy for peaceful purposes in the Warsaw Pact countries', Soviet Studies 26: 568-90.

8.   A. Dingsdale (1984) 'Hungary's energy crisis and nuclear power', Geography 69: 348-50.

9.   L. Gooz (1984) 'Utilization of alternative energy resources on the Great Hungarian Plain', in Compton and Pecsi (eds) 233-46.

Section II

SETTLEMENT PATTERNS

Chapter Five

SOCIALIST CITIES AND THEIR HOUSING POLICIES

Socialist economic development is tied up inextricably with the settlement pattern. Large industries must be located in cities where the necessary services are available and where workers can either be accommodated in housing estates or transported to the factories from surrounding villages. However the inherited urban structure has not been conducive to a balanced regional allocation of investment. It was based essentially on the imperial system of the last century and much of Eastern Europe was peripheral to the great cities of Berlin, Istanbul, and Vienna (Figure 5.1). The achievement of independence by some of the Balkan nations led to the rapid growth of Belgrade, Bucharest, and Sofia while the compromise worked out in 1867 between Germans and Hungarians within the Habsburg Empire provided the stimulus for the growth of Budapest with a number of prominent industrial districts related to transport by railway and river (Figure 5.2). During the inter-war years urbanization proceeded in the context of nation states but much of the growth occurred in the capital cities and in a small number of provincial centres. Accelerated urban growth has been a prominent feature of socialist development, given the emphasis on large-scale industry and the impossibility of providing sufficient accommodation by the evacuation of non-active households which was carried out in some cities in the early years of the revolution to relieve pressure on housing. However urban growth has been selective, according to planning requirements (usually with particular emphasis on the administrative centre of each region) and there have been ideological overtones to sustain low-cost development (with a preference for communal

Figure 5.1 The cities of Eastern Europe in the 1870s

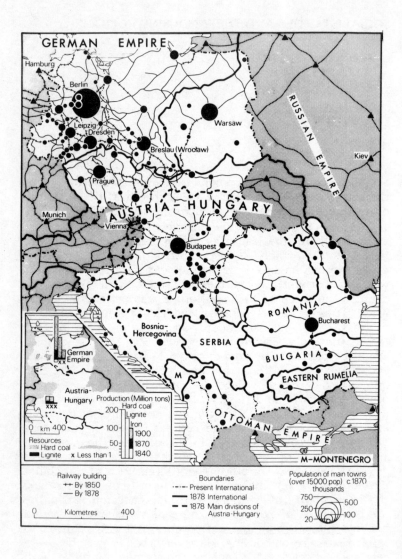

Source: National atlases.

Figure 5.2 Budapest: transport and industry

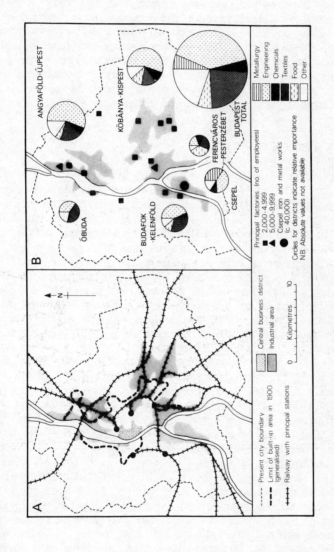

N

A

- ········ Present city boundary
- - - - Limit of built-up area in 1900 (generalised)
- +++++ Railway with principal stations

Central business district

Industrial area

0   Kilometres   10

B

ÓBUDA

ANGYAFÖLD-ÚJPEST

KŐBÁNYA-KISPEST

BUDAFOK-KELENFÖLD

FERENCVÁROS-PESTERZSÉBET

CSEPEL

BUDAPEST TOTAL

Metallurgy
Engineering
Chemicals
Textiles
Food
Other

Principal factories (no. of employees)
■ 2,000-4,999
▲ 5,000-9,999
● Csepel iron and metal works (c. 40,000)

Circles for districts indicate relative importance
NB Absolute values not available

Source: Bencze and Tajti 1972.

facilities over the privacy of the individual) and heighten worker motivation without the incentive of high wages. Studies are made of the urban and rural components of the settlement pattern, with initial emphasis on the former being tempered by greater concern in the 1980s for the villages. This is to ensure the economical servicing of a dwindling population in the countryside and to create a more viable base for the future rural economy which will involve more industry and tourism than before. The control of pollution through appropriate conservation policies is also discussed. But these studies are prefaced by a review of settlement planning and housing policies.

TOWN AND COUNTRY

The most significant redistribution of population has occurred through the growth of the towns at the expense of the countryside. (1) The urban areas are attractive because most new jobs in industry are available there, while mechanization results in fewer jobs in agriculture in the rural areas. (2) At the same time the organization of agriculture on a large-scale basis has reduced the need for the highly dispersed pattern of rural settlement inherited from the past. Communist governments have always looked positively at urban development as a means of combining a modern economy with a classless society offering equal opportunity for all citizens: urban living is the highest form of socialist life. (3) But practice has lagged a long way behind theory because insufficient resources have been allocated to housing and infrastructure. Priority has been given to investment in production and so concentration of production has gone ahead of concentration of population. Migration has been heavy but the housing shortage has restrained the flow to some extent and there are some incentives to remain in the country (good public transport for commuters and easy access to private agricultural plots)´ There may also be outright prohibition of migration to certain towns (except in special circumstances such as the need for key workers). So there is a great deal of daily commuting and Eastern Europe may therefore be considered under-urbanized in relation to the level of economic development. (4) But there have been no significant checks imposed for ideological reasons and in all countries there is an expectation of even higher urbanization rates by the end of the century: natural increase will be supplemented by in-migration from the countryside and while the existing towns will grow larger the number of towns will also increase through the promotion of villages to urban status. (5)

This is not to say that the authorities have never been ambivalent towards urban development. Early post-war notions of equality envisaged better standards of living in the rural areas through some decentralization of industry and the building of blast furnaces at Calbe on the Saale north of Bernburg in the GDR is an outstanding example. However with improvements in transport there has been a growing acceptance of the town as an efficient location for industry and as a provider of services for an extensive hinterland. Equality is therefore to be achieved by increasing interaction between town and country and an expansion of the complementary functions of each milieu within each district. Industrial developments in the towns therefore induce closer linkages between town and country. Wages taken back to villages mean new houses in the countryside and a more urban outlook in the villages as former peasant communities become less homogenous. Physical and functional linkages between settlements can only increase as industrialization proceeds and the process may be guided by location decisions over specific major industries. Work in the GDR has examined the varied demographic trends in rural communes in the context of access to major centres (agglomeration cores), recognizing the limits of daily commuting as an important means of discriminating between an inner zone of relatively intense interaction (agglomeration area) and an outer zone (agglomeration peripheral area) where population is more likely to be in decline with primary industries much more dominant. (6) However to maintain a balance between town and country the rural dwellers should be stimulated by better local living conditions and not simply be influenced by migration controls.

With some of the more difficult transition years now passed the majority of rural households now enjoy a regular income with reasonable access to employment and services either in towns or in key villages. The housing shortage in the towns, the benefits of a link with the land (especially in the Balkans where the climatic conditions boost the potential of gardens) and the state's concern to improve rural conditions in order to retain an adequate work-force for modern farming all contribute to this situation.

Table 5.1 Urban population of Eastern Europe 1950-80

Country	Urban population (millions) and share of total population (per cent)								Annual growth of urban population (per cent)		
	1950		1960		1970		1980		1950-60	1960-70	1970-80
Albania	0.25	20.5	0.51	30.6	0.80	37.4	0.90	33.6	0.32	5.75	1.20
Bulgaria	2.00	27.5	3.01	38.2	4.51	52.9	5.51	62.1	5.02	5.01	2.21
Czechoslovakia	6.35	51.5	7.89	57.5	8.94	62.3	10.16	66.4	2.41	1.35	1.36
GDR	13.04	72.0	12.37	71.7	12.59	73.8	12.75	76.2	-0.52	0.15	0.12
Hungary	3.55	38.6	4.34	43.5	4.99	48.2	5.70	53.2	1.54	1.20	1.42
Poland	9.61	39.0	14.40	48.7	17.09	52.3	20.29	56.8	4.99	1.83	1.87
Romania	3.71	23.4	5.99	32.5	8.34	40.9	11.01	49.6	5.92	3.97	3.20
Yugoslavia	3.27	21.9	5.24	28.3	7.38	35.9	10.00e	44.8	6.03	4.09	3.55
Eastern Europe	41.79	39.4	53.74	45.2	64.64	51.2	83.32	63.1	2.85	2.00	2.89

Sources: French and Hamilton (1979) 168-71; statistical yearbooks.

Note: e Estimate

Governments are starting to draw up rural development plans which take note of local economic resources (mainly agricultural but sometimes woodlands and minerals are important) and the potential arising from the population age structure and position in the context of the city region. Some governments are anxious to consolidate rural settlement in key villages where a full range of services can be provided and where cheap building plots might act as a further stimulant to abandon the smaller and remoter places. But others seem content with a laissez-faire approach. In the GDR outlying settlements are supported by 'transitional arrangements' like travelling shops and surgeries. It is recognized that many are likely to be of long-term interest for retirement and recreation despite the scattered nature of housing first built for tenant farmers. (7)

URBAN GROWTH: NORTH AND SOUTH

Migration has been taking place constantly during the post-war period. (8) In Poland alone 6.41 million people have migrated between 1951 and 1978: 0.23 million per annum. When allowance is made for the growth of population the change in actual rates from 0.28 million per annum in the early 1950s to 0.31 in the late 1970s is not significant. But the numbers moving from country to town have increased sharply, from 0.72 million per annum in the early 1950s (25.7 per cent of all migration) to 1.12 in the late 1970s (36.3). Put another way the net rural-urban migration of 0.87 million in the 1950s (equivalent to 6.2 per cent of the total urban population at the end of the decade) rose to 1.20 million in the 1960s (7.1 per cent) and 2.01 in the 1970s (9.4). The urban population is increasing (the only decline occurring in the GDR in the 1950s) and the rate of increase is greater than the rate of growth of population as a whole, enabling the urban sector to increase as a proportion (Table 5.1). There is a very strong contrast between the four Balkan countries with consistently above-average rates of urban growth and the northern countries. However, urban growth rates exceed the average in the 'recovered territories' of Poland and also in the eastern regions of that country; also in Slovakia and parts of Bohemia and Moravia (Czechoslovakia). The north was already highly urbanized on the eve of the Second World War: 45.4 per cent compared

157

Table 5.2 Rural and urban population trends in the GDR and Romania 1948-80

Country/category		1948		1956		1970		1980	
GDR		18,503		17,181		17,043		16,737	
Rural		5,463		4,755		4,424		3,988	
Urban		13,040		12,426		12,619		12,749	
Small towns		6,011		5,303		4,939		4,615	
Medium towns		2,329	(68)	2,459	(76)	2,633	(83)	2,468	(82)
Large towns:	Total	4,700	(23)	4,664	(24)	5,047	(29)	5,666	(35)
	50,000-100,000	877	(12)	961	(13)	1,081	(16)	1,318	(20)
	100,000-1 million	2,633	(10)	2,631	(10)	2,876	(12)	3,190	(14)
	Above 1 million	1,190	(1)	1,072	(1)	1,090	(1)	1,158	(1)
Romania		15,873		17,489		20,252		22,201	
Rural		12,160		12,742		11,994		11,187	
Urban		3,713		4,747		8,258		11,014	
Small towns		992		1,321		2,679		2,432	
Medium towns		607	(21)	738	(25)	1,256	(48)	1,626	(54)
Large towns:	Total	2,114	(13)	2,688	(16)	4,323	(25)	6,956	(40)
	50,000-100,000	842	(10)	653	(8)	850	(12)	1,460	(21)
	100,000-1 million	230	(2)	857	(7)	1,998	(12)	3,635	(18)
	Above 1 million	1,042	(1)	1,178	(1)	1,475	(1)	1,861	(1)

(The four year groups fall under the heading **Population**.)

Source: Statistical yearbooks.
Notes: Numbers in parentheses refer to the number of towns in the relevant categories.
Thresholds for the urban classification are 20,000 and 50,000.

with 20.03 in the Balkans. The north then held 81.0 per cent of the urban population of Eastern Europe. Indeed urbanization rates remain extremely high in a belt extending from the southern parts of the GDR and northern Bohemia south-eastwards to Silesia and northern Moravia, reflecting the major industrial areas of the capitalist period. But by the end of the 1970s the urbanization rate in the north had advanced to 58.6 per cent while the rate in the south doubled to 40.7: so the north then accounted for only 67.5 per cent of the total urban population of Eastern Europe.

The contrasts are most fully revealed in a comparison between the GDR and Romania (Table 5.2). In 1950 the GDR's population was already highly urbanized (72.0 per cent compared with only 23.4 per cent for Romania) and since then Romania has experienced a substantial overall increase of population (36.1 per cent) compared with a decline of 6.7 in the GDR. The growth of the urban population since 1950 has been much more rapid in Romania than in the GDR (296.8 per cent as against 2.2) but it is interesting to note that in 1980 the GDR was still substantially more highly urbanized (76.2 per cent compared with 49.6). There were 6.9 towns with a population exceeding 20,000 for every million of the total population compared with 3.3 in Romania. Further analysis in Table 5.3 introduces a rather longer time scale and indicates the trends in respect of nine different size groups to which towns are assigned with regard to their population in 1950. It is evident that whereas between 1905 and 1950 (1910 and 1948 in Romania) growth rates tended to be more rapid in the larger towns, after 1950 there has been less variation between the groups. It seems clear that post-war planning has not resulted in a perpetuation of the inherited hierarchy. When the performances of individual towns are considered for the period 1950-80 (taking into consideration those towns with at least 15,000 inhabitants in 1950 and other towns which subsequently crossed this threshold) it emerges that in every size group there are some Romanian towns which have increased by more than 50,000 and there are examples of increases of more than 100,000 in more than half the categories (Table 5.4). The picture for the GDR is even more irregular. While numerous cases of decline might be expected (against none in Romania) it is surprising to find four examples relating to cities with a 1950 population exceeding 75,000 (Gorlitz, Leipzig, Plauen, and Zwickau). Where growth is concerned it is noticeable that 59 of the 72

159

Table 5.3 Urban population trends in the GDR and Romania 1905-80 by size groups for 1950

Country/size group		Total population (thousands)					Percentage population change				
		1905	1939	1950	1965	1980	1905-39	1939-80	1939-50	1950-65	1965-80
GDR	139	5,315.3e	8,092.1	7,837.4	7,842.0	8,422.0	+52.2	+4.1	-3.1	+6.0	+6.0
15,000-20,000	42	476.3e	634.2	739.6	764.4	803.3	+33.2	+26.7	+16.6	+3.4	+5.1
20,000-30,000	31	465.3e	651.4	752.8	758.3	823.8	+40.0	+26.5	+15.6	+0.7	+8.6
30,000-40,000	25	511.9	754.1	854.3	859.8	845.2	+47.3	+12.1	+13.3	+0.6	-1.7
40,000-50,000	18	525.0	732.2	823.7	813.3	830.9	+39.5	+13.5	+12.5	-1.3	+2.2
50,000-75,000	6	246.1	365.7	345.7	375.5	443.2	+48.6	+21.2	-5.5	+8.6	+18.0
75,000-100,000	6	327.3	532.3	531.1	553.7	629.1	+62.6	+18.2	-0.3	+4.2	+13.6
100,000-200,000	5	371.7	601.7	678.8	704.0	781.3	+61.9	+29.8	+12.8	+3.7	+11.0
200,000-1 million	5	1,644.7e	2,232.2	1,952.6	1,941.5	2,017.6	+35.7	-9.6	-12.5	-0.6	+3.9
Above 1 million	1	747.0e	1,588.3	1,158.8	1,071.5	1,157.6	+112.6	-27.1	-27.0	-7.5	+8.0
Romania[a]	51	1,731.3	2,332.7	3,030.6	4,595.9	7,462.2	+34.7	+219.9	+29.9	+51.6	+62.4
15,000-20,000	16	222.3	241.7	272.7	394.5	706.3	+8.7	+192.2	+12.8	+44.7	+79.0
20,000-30,000	14	242.7	281.0	334.7	573.4	970.7	+15.8	+245.4	+19.2	+71.3	+69.3
30,000-40,000	4	98.9	97.3	121.6	185.3	372.0	-1.6	+282.3	+25.0	+52.3	+100.8
40,000-50,000	4	109.5	156.8	178.8	293.6	529.3	+43.2	+237.6	+14.0	+64.2	+80.3
50,000-75,000	1	33.5	49.3	67.6	109.7	161.1	+47.2	+232.9	+37.1	+62.3	+49.6
75,000-100,000	9	524.5	688.2	790.8	1,312.8	2,215.4	+31.2	+221.9	+14.9	+66.0	+68.8
100,000-250,000	2	133.4	192.4	246.1	359.8	578.0	+44.2	+200.4	+27.9	+46.2	+60.6
Above 1 million	1	366.5	626.0	1,018.3	1,366.7	1,929.4	+70.8	+208.2	+62.7	+34.2	+41.2

Source: Statistical yearbooks.
Notes: a   Census years are 1910, 1930, 1948, 1966, and 1981.
       e   Estimate

cases of growth in excess of 2,500 involve increases of less than 25,000, a higher proportion than in Romania, and that the 13 cases of more substantial growth are drawn from a range of size groups including five towns with a population below 30,000 in 1950 (Eisenhuttenstadt, Hoyerswerda, Neubrandenburg, Suhl, and Wolfen). This reflects the high level of discrimination between towns in the allocation of investment with disproportionate increases occurring in some new towns and in other places recognized as administrative centres for their regions.

SOCIALIST CITIES

A fundamental question which may appropriately be asked in connection with the large cities of Eastern Europe concerns their distinction as 'socialist cities'. (9) Whole volumes have been written on such issues but no clear consensus has emerged. Very simply it could be argued that because we are dealing with socialist countries the towns can be regarded as 'cities of socialism'. However this is not to say that they have characteristics which make them fundamentally different from western cities for all time. Different approaches may be taken. First there is the question of the urban network and here it would appear that after the aberrations of the early post-war years there is now a close similarity in that a hierarchy of urban influences now exists. Initially investment was relatively concentrated and although Rugg thinks there is some evidence of dispersal of industry for ideological reasons it transpires from his examination of early post-war Eastern Europe that the initial autarkic phase saw an emphasis on the leading cities. Generally speaking the small towns did badly in the early post-war years, because of reduced trading functions at a time of nationalization of small businesses and compulsory deliveries in agriculture, but experienced considerable growth after the 1950s when the planners began to exploit their labour catchments. Settlement policies have worked for greater national integration through a hierarchical system of urban settlements, seeking a balance between economic efficiency and social equity. This would appear to be working along conventional western lines. The increasing demand for urban services far away from the larger towns increases the importance of lower-order centres and some towns in

161

Table 5.4 Changes in the population of individual towns in the GDR and Romania 1950–80

Column groups — **Towns in group:** A, B, C · **Decline:** D, E · **Decline or growth in individual towns where greater than 2,500 (Growth):** F, G, H, J, K, L. Each lettered cell (except A and B) shows a count followed by a value.

Country/size group (1950)	A	B	C	D	E	F	G	H	J	K	L
GDR	139	41	7 132.4	10 64.2	26 94.3	20 73.7	18 129.4	21 339.6	9 279.6	3 174.4	1 101.4
Below 15,000	42	17			11 37.2	6 23.6	3 22.5	5 75.9	2 74.5	1 63.2	
15,000–20,000	31	12			7 24.1	7 26.1	3 20.8	4 53.4			
20,000–30,000	25	5	2 20.3	4 23.2	7 27.2	1 11.5	1 7.0	3 10.9	1 25.1	1 57.3	
30,000–40,000	18	3	2 20.3	5 34.3		2 4.4	2 16.9	3 50.9			
40,000–50,000	6	3		1 6.7		2 8.1	9 62.2	1 17.1			
50,000–75,000	6	0						1 16.8	1 27.9	1 53.9	
75,000–100,000	5	0			1 5.8			2 23.7	3 80.1		
100,000–250,000	5	0	2 36.1					2 42.2			
250,000–1 million	5	0	1 55.7					2 48.7	2 72.0		
Above 1 million	1	1									1 101.4
Romania	51	0					22 180.9	52 814.3	22 795.4	17 1,187.9	13 2,809.2
Below 15,000	16	0					20 164.3	41 532.3	12 430.6	2 137.3	
15,000–20,000	14	0					2 16.6	8 126.0	3 104.4	3 177.6	
20,000–30,000	4	0						3 56.0	6 233.7	5 346.3	
30,000–40,000	4	0							1 26.7	3 110.2	
40,000–50,000	1	0								3 234.5	1 112.5
50,000–75,000	9	0								1 96.5	1 116.0
75,000–100,000	2	0								1 85.5	8 1,338.2
100,000–250,000	2	0									2 331.4
250,000–1 million	0	0									
Above 1 million	1	0									1 911.1

Source: Statistical yearbooks.
Notes: Only those towns with more than 15,000 population in 1980 are considered. Account is taken of towns which crossed the threshold
between 1950 and 1980, through the 'Below 15,000' (in 1950) size group, although figures for columns A and B are not available.

A   Total number
B   Number with decline/growth not exceeding 2,500
C   Greater than 10,000
D   5,000-9,999
E   2,500-4,999   decline
F   2,500-4,999
G   5,000-9,999
H   10,000-24,999
J   25,000-49,999   growth
K   50,000-99,999
L   100,000 and over

formerly backward areas have shown spectacular growth, for example Blagoevgrad in Pirin Macedonia, an area which the Bulgarians are particularly anxious to develop in order to counter accusations of neglect from Yugoslavia. However, some small towns continue to lose ground, especially where there are large cities close by, enabling villages to establish direct relations with the higher order centres. The larger centres grow while the hinterlands of the smaller towns collapse, although the latter may find new functions as dormitories for the wider conurbation. (10) Thus the cathedral town of Kalocsa in Hungary remains an agricultural community with paprika cultivated on a large scale in the surrounding area.

Morphology

Consideration may be given to urban layouts but it is noticeable that there is no distinctive model form for the socialist city. The concept of the linear city evident in the work of Soria y Mata and Le Corbusier was applied to some Soviet industrial settlements in the 1930s. The idea was championed by the Soviet planner Milyutin and put into practice at Stalingrad (now Volgograd). The city expanded in linear fashion along the river bank through a chain of residential complexes linking with industrial zones situated on the other side of the green belt. Similar tendencies are evident in the growth of other Soviet cities along river valleys but the model seems to have lost much of its clarity and there is no obvious impact in Eastern Europe. New housing may be concentrated along the major thoroughfares but such linear developments radiate from the centre in all directions and there is no binding relationship between one housing complex and the industries which have emerged in the same neighbourhood: complex travel to work patterns include many cross-city journeys. Some socialist idealism has focused on the neighbourhood concept which conflicted with the economic efficiency of a band of development along major transport axes. Socialist planners have tried to emphasize the individual neighbourhood as an environment for living and working, with facilities for shopping, schooling, and cultural pursuits all available locally. But this is not an intrinsically socialist conception and planners are well aware of urban development practice in capitalist countries. The initial Soviet emphasis on communal living, reflected in the concept of 'mikrorayon' elaborated by Strumilin, has given way to more conventional arrangements involving self-contained family apartments. Furthermore while socialist cities are conventional in expanding both upwards (with high-rise apartment blocks) and outwards they also reflect orthodox 'garden city' thinking through the development of new towns as decongestion points, appearing individually or in groups to generate an urban agglomeration. Thus the new Moravian mining town of Havirov is closely integrated with the city of Ostrava. Existing small towns also contribute to the formation of urban clusters focusing on a large city with green spaces interposed to prevent continuous urban sprawl.

But over functional specialization there appear to be different approaches. 'In Sofia the accent is on functions for each separate zone whereas Prague will have independent sectors with a self-contained urban economy. (11) Basically there are six sectors each with its own facilities. The hierarchy then extends down to districts within the sectors and finally to local centres with their own food shops, catering facilities, and other basic services. Some Polish research reveals the primacy of the 'central shopping area' in the local trading hierarchy and also identifies some suburban districts as having 'high trading attractiveness' which could in future justify a polycentric arrangement with some of the central shopping functions diverted to the suburbs. (12) And despite the ideal of suburban housing schemes integrated with industry and services there is a great deal of commuting within the city, including lengthy cross-city flows. In Lodz it has been shown that the commuting pattern has much to do with the historical evolution in the pre-socialist period. (13) Moreover the various neighbourhoods appear to have different status levels. Dealing with Warsaw Weclawowicz demonstrates that in spite of reconstruction and a wholesale change in the population the city retains a mosaic spatial structure and is not completely homogenous. 'Despite many restrictions the operation of economic factors of spatial selection should not be under-estimated in fostering and consolidating spatial structures.' (14) Indeed the mosaic structure will probably be stabilized as a typical pattern for a socialist city. Again despite varied occupational status of residents in housing estates there are obvious contrasts in amenity between districts of cities. (15)

Architecture

Urban architecture has also been scrutinized as a relevant criterion. The expectation is that whole towns or certain neighbourhoods will appear outwardly as socialist creations through the emergence of buildings of a suitably monumental character. The authorities may wish to take advantage of an opportunity arising through the need to repair wartime damage in areas previously lacking in coherence or deserving of recognition through a recent history of worker unrest. The Grivita district of Bucharest has been seen as a particularly striking example of revolution in townscape art. Grivita is the district in which the main railway station is situated, along with administration, locomotive and carriage sheds, and repair shops. The area is noted for the strike which broke out in the railway workshops under the leadership of Gheorghe Gheorghiu, a man who was subsequently transferred to the small Transylvanian town of Dej but only to rise in the ranks of the Communist Party and emerge as the post-war leader Gh. Gheorghiu-Dej. In view of the wartime bombing and the scope for development there was scope for a powerful architectural effort to create a socialist gestalt. The Bucharest Studio for the Plastic Arts of the Army turned out the blueprints for monuments comprising purposeful human figures complete with tensed arms, clenched fists, and sober faces. Yet in Grivita today this work plays only a minor role in an architectural ensemble which is essentially pre-communist, as evidenced by the new railway terminus building and the administrative building which date back to the 1930s. Some old buildings also retain their relevance in socialist society.

> The situation of many of the city's old churches suggests that they serve as constrasting visual foci in many recent construction projects. In addition they imply the existence of the 'bisericuta' (a term deriving from the word for 'parish', now used to refer to a circle of friends) in the new gestalt, while at the same time continuing as an artistic feature of the capitalist urban image. (16)

However all this is plainly irrelevant to the present planning of Bucharest with an uncompromising ethos of modernization which rejects more of the traditional elements. Many churches have been demolished over the last decade and those saved through international protest have been moved aside on steel platforms so that the new facade is in no way interrupted.

What seems to emerge therefore (at least for the 1950s and 1960s) is the understandable desire of the party to commemorate the liberation from enemy occupation and the frustrated ambitions of capitalism with appropriate architectural manifestations. But the shortage of capital for non-productive building means that the scale of such work must be modest and it cannot greatly modify the inherited urban landscapes which project nationalist traditions. Indeed communist governments have been forced to compromise 'between revolutionising the feudal or capitalistic society fossilised in stone while sustaining pride in national heritage and devoting the utmost resources to ambitious plans for economic development and socialist transformation. (17) As far as building in general is concerned it may be most appropriate to see a reassertion of national characteristics, albeit in the context of mass production techniques and high-rise developments (Wohnscheiben), following the early post-war phase of conservative formalism with heavy ornamented styles, especially the Soviet-inspired 'wedding cake' style (Zuckerbäckerstil). Conservation work also projects national traditions, all the more so in view of attempts to harmonize new buildings with the inherited fabric. Immediately after the Second World War the rebuilding of Warsaw involved the reconstruction of the 'old town' (as well as the appearance of a massive Stalinist edifice for the Palace of Culture) and the cores of Gdansk, Lublin, and Poznan have also been restored. Evidence can be found from the other countries of Eastern Europe with additional pressures to conserve historic townships evident in the south through the development of tourism (cities like Sibiu in Romania and Plovdiv in Bulgaria). Several conservation areas have now been recognized by UNESCO as being of world significance: the centre of Krakow (and the Wieliczka salt mine) in Poland and the Yugoslav towns of Dubrovnik, Kotor, and Split. These historic cities are certainly quite different from the new towns like Dimitrovgrad, Dunaujvaros, and Eisenhuttenstadt.

The ideological factor

It has been asserted that there is a significant ideological factor which is more than a hollow ritual. Equal opportunity is the key: 'irrespective of location, ethnicity, skills or income all people should have access to the same standards or norms in housing, transport, education, medical care and cultural/recreational facilities'. (18) However the search for equity is not a monopoly of communist societies and the fact that punitive action has been taken against the wealthy, through expropriation of property and erosion of cash savings through currency reform, should not give rise to an assumption that all inequalities have been eliminated. There is much evidence to suggest continuing inequality as regards the quality of life between different settlements and within individual towns and cities. For example contrasts arise between well-maintained town houses available to the elite and the poorly serviced rural-type properties. Again the introduction of central planning of the economy does not necessarily promote equality. Planning at the regional and local levels is constrained by the wide-ranging powers of the ministries whose investment decisions are crucial as far as growth in individual cities is concerned. Only in 1961 did planners in the Polish regional administrations have any power to refuse developments sought by ministries. Then in 1975 there was a further strengthening of the regional planning system as forty-nine voivodships were set up with co-ordination of plans at the level of the macro-region. There are eight of these in Poland, focusing on Bialystok, Gdansk, Katowice, Krakow, Lublin, Poznan, Warsaw, and Wroclaw. Plans for individual settlements are to involve People's Councils as fully as possible thereby increasing public interest and approval. Thus it may be questioned how far there has been planned urban development in Eastern Europe because the emphasis on central planning has meant that 'the ability of the authorities at the local level to oblige the managers of state-owned enterprises to agree to, let alone conform with, long term local plans for town development has been limited'. (19)

Moreover, the urban planning that has been attempted has not always been consistent and any underlying principles and models have not been projected clearly. 'Frequent political changes within the communist party have been reflected in continual changes in the implementation of plans and the interpretation of socialist planning principles'. (20) The decision to build the Nova Huta quarter in Krakow and then to extend that new suburb well beyond the initial spatial and demographic limits provides a clear example. Housing provision has suffered from the early post-war attitude that such a preoccupation is economically unproductive and at best a necessary evil (as in the case of capital cities where there are important propaganda considerations) and certain suburban developments in provincial cities where there are important political issues involved, as at Nova Huta. Industrialization has been strongly encouraged rather than the development of broadly based urban economies: compared with other countries with similar levels of GNP per capita East European countries tend to be high on employment in industry but low on employment in services. East European scholars are quite unequivocal about the basic realities. The Polish geographer G. Weclawowicz asserts that 'a characteristic feature of socialist countries is the domination of politics over the economy' and that at the level of the city 'the concentration of expenditure on the expansion of industry resulted in an insufficient expansion of the building industry in relation to needs'. (21) The serious housing shortages, combined with inadequate attention to the renovation of old properties results in a large army of 'peasant workers', particularly numerous in areas like southern Poland. When productivity is blunted by the fatigue of long journeys to work planned economies of scale may well be converted into diseconomies (quite apart from the social costs to the individuals involved).

Such considerations produce distinctive urban conditions which are the reverse of what ideological principles would suggest. The limited importance of the tertiary sector means that city centres are not dominated by shopping and other service functions. Rather there is a substantial permanent population with many buildings providing apartments on all the floors above ground level. This is a common phenomenon in old property and also in new blocks which have emerged on bombed sites or on large areas cleared to make way for a new town centre. The mixing of functions in the centre is further encouraged by the lack of any selection mechanism based on land-use competition working through land and property values. State or co-operative control of retailing and much of the housing also simplifies the development process by placing it in the

169

hands of the planners and the politicians behind them. Land is being regarded as an increasingly valuable resource but the value is expressed not so much in a surface of land values as by arbitrary density standards which seek to minimize land taken from agriculture on the edge of cities to make way for new housing complexes. Restrictions on the amount of living space available to each person prevent major disparities between owner-occupied and rented accommod-ation and in Romania the effective prohibition of all cottage-type development in cities means that all new developments comprise high-rise blocks with only minimal architectural variety. This also reduces investment in infrastructure since a modest extension to the tramway system and the networks of gas and water mains can serve a large number of additional families. In the new suburban 'white belts' densities may exceed 11,000 per square kilometre, usually higher than values encountered in the core, so that the density gradient phenomenon for socialist cities may sometimes be the reverse of the normal western situation. There are close similarities in architecture and also in the ethnic and occupational composition of the tenants (with relatively little segregation by western standards). The low level of car ownership results in a low level of garage provision: cars do however clutter the streets and those not in regular use may be stored under waterproof covers. By contrast public transport is very prominent and depots for buses, trolley-buses, and trams are therefore prominent. Finally, given the political power of the communist parties there is an emphasis on party propaganda rather than private advertising and on the provision of communal facilities rather than extended living spaces for individual families.

It is a matter of opinion how far such characteristics reflect communist ideology as opposed to national characteristics or economic stringency. National character would not appear to be very relevant here, except in matters of detail such as the prominence of certain types of shops like the cafes in Polish cities and the 'slasticarne' in Yugoslavia, the prominence of some architectural details (like the Brancovan arches included in many Romanian buildings), and the use of distinctive systems of housing administration like the tradition of co-operative housing in Bulgaria involving three to five storey buildings constructed jointly by families to whom the apartments were eventually allocated as owners. The system worked well in the inter-war years and is still in use, as a national style, complementing the large estate developments by the government and the municipalities. For the rest it may be argued that both ideology and scarce resources conspire to limit private car ownership and large houses or apartments. But in the future will ideological principles create widening disparities between communist and capitalist cities or will greater prosperity allow higher living standards, greater individualism, and a general narrowing of the differences? Again, it is clear that there is much more homogeneity on housing estates than would be expected in the west. But of course research of this kind must be taken in context: housing shortages combined with low wage levels prevent the same exercise of choice that is the privilege of most families in the west, while it is possible that the greater number of informal measures for dealing with anti-social elements reduces pressure for segregation. And does the lack of a land value surface reflect a firm ideological position or a rather unsophisticated approach to planning which may change in future?

The ideologists like Hamilton argue that

the ability to achieve socialist objectives in managing specific agglomerations depends upon the stage in their development at which effective urban spatial planning is introduced . . . Policies have greater chances for success in shaping the embryonic agglomerations . . . or emerging agglomerations of bigger towns such as Bydgoszcz-Torun (Poland). (22)

By contrast difficulties are likely to arise in agglomerations that were already very extensive in the 1950s and especially in those areas where there are serious problems of congestion and decayed housing as in such Polish cities as Katowice and Lodz. For the townscape will clearly relate significantly to the capitalist past rather than the socialist present and despite some efforts at redevelopment the emphasis will necessarily rest on adaptation of old property to serve the needs of the new society. Conceding that housing is certainly differentiated in quality within and between cities Hamilton asserts that this is a legacy of the pre-socialist past and is the fault of post-war socialist planning 'only insofar as replacement rates of older by modern uniform housing stock has been insufficient to eliminate the significant role that the market in

171

privately-owned older dwellings plays in the social distribution of shelter'. (23) Rugg claims that the 'landscape has a new element under socialism that cannot be subsumed by merely calling it modernisation' so that there is 'a degree of incompatibility between the traditional forms and those of socialism'. (24) Presumably he means that equity goals carry a far higher priority in Eastern Europe than they do in the west. But on the other hand Dawson considers that the differences between socialist and western cities will diminish through 'the common adoption of improved methods of transport, traffic control, building and planning'. (25) And in Budapest Compton concludes 'although urban processes are occurring within the framework of a planned socialist system they appear to differ little from those operating in Western cities'. (26)

The issue is complex and clarification would seem to require greater definition. In particular the ideologists should identify those elements of socialist cities which reflect equality of opportunity as sought by both the communist authorities and the people who inhabit the cities they create. If ideology is not to be a hollow ritual it must generate improvements over capitalist cities which are perceived as such by the mass of the population as well as the ruling elite. R. Bahro would argue that although the communist movement emerged in a bid to solve the basic problems of modern humanity it has not attained its promised breakthrough: 'for the time being we are continuing on the capitalist road: the alienation of the working masses persists in a new phase'. (27)

THE HOUSING PROBLEM

Before the Second World War the Czechoslovak Communist Party demanded decent housing for everyone but after more than three decades of absolute rule the party has not succeeded, although it started with a housing fund well stocked by the acquisition of homes of some 3 million Sudeten Germans expelled in 1945 (not to mention tens of thousands of exiles who left between 1948 and 1968). The number of new dwellings built in Czechoslovakia rose from 30.5 thousand per annum between 1946 and 1955 to 72.2 thousand over the next ten years and 105.4 thousand between 1966 and 1975. The shortages continue, exacerbated by the internal migration which left almost half a million rural houses abandoned between 1961 and 1975: this cancels out half the new housing built in the towns during the same period. The solution of the housing problem has been promised before 1970 and then before 1985, but to no avail. According to a long-term Polish housing programme in 1972 every family was to be assured of its own home by 1990 at the latest. The increased purchasing power in the 1970s resulted in a new housing drive which was meant to provide a solution to housing shortage by the late 1980s. But the number of households per 100 dwellings fell less between 1970 and 1981 (116 to 115) than in the 1950s (122 to 118) or 1960s. The shortage is now destined to continue for another two decades and differences in the general quality of housing between the 'recovered territories' and the 'former' areas and between urban and rural areas will also persist. (28) Some 1.1 million new apartments are to be built between 1986 and 1990, increasing the total area of settlements by more than 10 per cent. It remains to be seen if this ambitious target can be met.

Table 5.5 shows the trends in house-building in Romania as a whole from 1951 to 1985. When building is related to the population level a very rapid improvement can be seen in the late 1950s because the rate increased from 3.12 houses per thousand of the population in 1950 and 3.20 in 1955 and 7.27 in 1960. The rate fluctuated over the next twenty years between 5.76 in 1968 and 8.91 in 1980, but it is noticeable that since 1980 there has been a steady reduction in effort in both town and country down to the level of 4.65 in 1985, the lowest rate since the 1950s. Major differences occur in the contributions by the state and by private investors. Private building was much more important than the state programmes in the 1950s but the state sector became the larger of the two in 1969 and the gap has widened since, with the state contribution at least ten times larger than the private sector during the last seven years. The reasons lie primarily in the party's decision to become more involved in housing so that apartments have become available for renting or purchase on very attractive terms. This has been especially true in the larger towns. At the same time urban planning has focused more and more narrowly on the apartment block and so it has become increasingly difficult for plots to be found on which new private houses can be built. It should however be stressed that the figures relate only to green-field development.

173

Table 5.5 House-building in Romania 1951–85

Milieu	Period	State sector A	B	C	Private sector A	B	C	Total A	B	C
Romania	1951-55	9.3	26.0	0.53	77.3	26.5	4.42	86.6	26.5	4.95
	1956-60	20.8	29.7	1.13	151.4	26.7	8.23	172.1	27.0	9.35
	1961-65	44.0	30.7	2.31	137.1	25.5	7.21	181.1	26.8	9.52
	1966-70	66.6	29.1	3.49	62.9	34.9	3.29	129.5	31.9	6.78
	1971-75	102.5	29.8	4.82	47.9	40.1	2.25	150.4	33.0	7.08
	1976-80	151.1	32.6	6.80	17.1	49.5	0.77	168.1	34.3	7.57
	1980-85	131.1	33.6	5.79	10.2	52.4	0.45	141.3	34.9	6.25
Urban	1951-55	7.0	26.8	1.27	11.8	25.9	2.15	18.7	26.3	3.42
	1956-60	18.6	30.3	3.14	35.3	26.8	5.98	53.9	28.0	9.71
	1961-65	41.8	30.9	6.51	28.0	26.2	4.37	69.8	29.0	10.88
	1966-70	65.1	29.0	8.91	12.3	35.1	1.68	77.4	30.0	10.59
	1971-75	100.4	29.8	10.92	12.1	42.2	1.32	112.5	31.1	12.25
	1976-80	146.9	32.6	13.33	4.2	53.2	0.39	151.2	33.1	13.72
	1981-85	126.0	33.6	10.65	2.4	54.6	0.20	128.4	34.0	10.85
Rural	1951-55	2.3	23.7	0.19	65.6	26.7	5.46	67.9	26.6	5.65
	1956-60	2.2	25.4	0.18	116.0	26.6	9.29	118.2	26.6	9.47
	1961-65	2.3	27.4	0.18	109.1	25.3	8.65	111.3	25.3	8.83
	1966-70	1.5	31.6	0.13	50.6	34.8	4.29	52.1	34.7	4.42
	1971-75	2.1	29.5	0.18	35.8	39.3	2.96	37.9	38.8	3.14
	1976-80	4.2	33.7	0.37	12.8	48.3	1.15	17.0	44.7	1.52
	1981-85	5.1	33.0	0.47	7.9	52.0	0.73	12.9	44.6	1.20

Source: Statistical yearbooks.

Notes: A Houses built per annum (thousands)
B Average area (square metres)
C Houses per thousand of the population

They do not include major reconstruction of old houses (which may amount to almost total rebuilding), which is a major preoccupation among private builders in the rural areas. (29)

Table 5.5 also deals with the area of living space (suprafata locuibila) in new houses. It is clear that the trends are upwards from 26.5 square metres per house in 1951-5 to 34.9 in 1981-5 but the increase has been much more rapid in the rural areas (26.6 to 44.6) than in the towns (26.3 to 34.0); also in the private sector (26.5 to 52.4) compared with the state sector (26.0 to 33.6). Evidently the norm of 13 square metres per person is being increasingly exceeded, although many large privately owned houses accommodate extended families. The differences in building rates between town and country are particularly striking (Table 5.6). Whereas the urban rate was lower than the rural rate in 1951 (2.49 houses per thousand of the population as against 3.32), the urban rate overtook the rural in 1955 and it has stayed ahead ever since, apart from 1960 and 1961. The ratio has widened in recent years: 3.0:1 in 1970, 5.7:1 in 1975, 13.5:1 in 1980, and 19.2:1 in 1985. In the urban areas the growth of the state sector has been particularly impressive through the 1960s and 1970s while the private sector has registered an absolute decline since 1970. In 1985 45 houses were built by the state for each one built privately compared with 15 in 1975, 7 in 1965 and 1.5 in 1955. In the countryside private building is still prominent although the state sector is beginning to catch up, reaching two-thirds of the private sector level in 1985 compared with one-seventh in 1975 and one-twentieth in 1964. While private building on new sites is falling off, because it is administratively simpler to reconstruct existing properties, the state is acquiring land in many of the key villages for the construction of small apartment blocks.

A finer focus can be obtained by looking at the situation in individual regions (Figure 5.3). In this illustration building rates, state participation, and average house size are examined in terms of deviations from the national average picture for 1970-85 as a whole. And in the case of the first two criteria information is also given for 1970-4, 1975-9, 1980-4, and 1985. What comes out very clearly is the above-average building effort in those counties where the largest cities are found. These cities will have attracted migrants from well beyond the limits of the relevant county and therefore building rates will be high

Table 5.6 Annual housing returns for the urban and rural areas of Romania

Year	Urban sector A	B	C	Rural sector A	B	C	Total A	B	C
1951	1.34	1.15	2.49	0.07	3.25	3.32	0.36	2.76	3.12
1955	2.39	1.20	3.59	0.21	2.81	3.02	0.89	2.31	3.20
1960	4.63	1.72	6.35	0.22	7.49	7.71	1.63	5.64	7.27
1965	7.41	1.07	8.48	0.27	5.01	5.28	2.68	3.68	6.36
1970	11.50	1.98	13.48	0.12	3.87	3.99	4.76	3.10	7.86
1971	10.47	1.91	12.28	0.09	3.53	3.62	4.32	2.86	7.18
1972	9.85	1.53	11.38	0.10	3.06	3.16	4.16	2.42	6.58
1973	11.27	1.35	12.62	0.16	3.04	3.20	4.83	2.83	7.16
1974	11.63	1.24	12.87	0.23	2.99	3.22	5.10	2.24	7.34
1975	13.79	0.92	14.71	0.33	2.19	2.52	6.15	1.64	7.79
1976	11.56	0.69	12.25	0.31	1.70	2.01	5.24	1.26	6.50
1977	11.73	0.48	12.21	0.34	1.30	1.64	5.79	0.91	6.70
1978	13.93	0.40	14.33	0.33	0.96	1.29	6.94	0.69	7.63
1979	16.28	0.27	16.55	0.39	0.84	1.23	8.13	0.56	8.69
1980	16.48	0.24	16.72	0.46	0.76	1.22	8.40	0.51	8.91
1981	12.98	0.21	13.19	0.47	0.76	1.23	6.73	0.49	7.22
1982	12.54	0.22	12.76	0.46	0.78	1.24	6.68	0.49	7.17
1983	11.01	0.21	11.22	0.55	0.82	1.37	6.00	0.50	6.50
1984	9.94	0.20	10.14	0.44	0.66	1.10	5.41	0.42	5.83
1985	7.69	0.17	7.86	0.40	0.61	1.01	4.27	0.38	4.65

Source: Statistical yearbooks.

Notes: A Total number of new houses per thousand of the population in the state sector
B Total number of new houses per thousand of the population in the private sector
C Total

when related to the population of the county alone: Brasov +1.91, Cluj +1.61, Constanta +2.02 and Giurgiu-Ilfov (covering Bucharest) +3.98. It is likely that there is a link here with above-average levels of state participation in house-building since a large predominantly youthful immigrant population lacking any local family ties will require substantial assistance from the state building programme. However such a simple explanation cannot fit all cases and when the counties with the lowest levels of state participation are identified they form a block in the north of the country where the share of housing constructed by the state is below the national average by a substantial margin: Bistrita-Nasaud -9.8, Maramures -12.0, Salaj -12.6, Satu Mare -17.0, and Suceava -15.8. There is another grouping in the Subcarpathians in the south-east and south-west where the state contribution is relatively small: Buzau -9.6, Dimbovita -14.5, Gorj -11.8, Olt -11.4, and Vilcea -16.6. The northern counties are well known for a particularly cohesive community life in the villages where the population continues to increase (thanks to the availability of seasonal work in agriculture and forestry in other parts of the country). This expansion obviously creates a demand for additional house plots, whereas in the rest of the country, rural population is falling and demand for new housing can be satisfied more easily by the redevelopment of existing sites.

Urban housing complexes

Large housing complexes have resulted from new building technologies. Thus in Hungary an important beginning was made in 1965 with substantial investments in a network of factories producing prefabricated panels. Under an agreement with the USSR Hungary purchased nine Soviet factories producing prefabricated housing units and received Soviet construction technology. The national factory network was supplemented by one Danish factory. Hence the appearance of the tower buildings of eight to ten storeys: prefabricated units accounted for slightly less than a quarter of the million apartments built in Hungary between 1961 and 1975 and the proportion has risen since. At the SED congress in 1971 the initiative was taken to eliminate housing as a social problem by 1990. The plan will involve new housing complexes like Berlin-Marzahn (35,000 units),

Figure 5.3 Housing in Romania

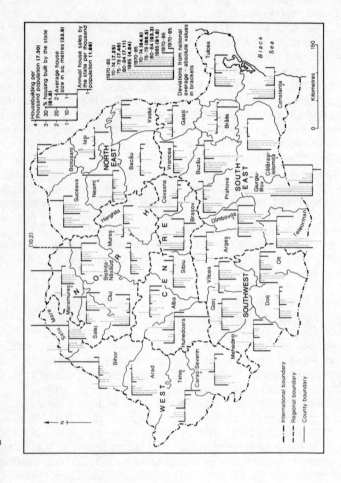

Source: Statistical yearbooks.

Leipzig-Grunau (20,000), and Magdeburg-Nord (15,000). This presumably involves attention to old properties which is a serious problem throughout Eastern Europe. Up to two-thirds of the housing stock built before 1945 must be replaced or renovated if housing, as a social problem, is to disappear. Renovations currently account for about one-third of all units provided. Increasing this proportion will be difficult because of labour demands (a higher proportion of craftsmen to labourers than is the case for the assembly of prefabricated units. The situation has been complicated by the very low rents fixed by the state for tenants living in private houses (for example the large traditional town houses now divided into flats). Lack of resources to keep the property in repair leads to progressive decay sometimes forcing the owners to beg the authorities to take over the responsibility. The GDR has also been improving housing conditions for chemical workers by rehousing families living close to the Buna and Leuna complexes (situated on the Saale south of Halle where lignite could be drawn from the Giesel valley) in the new suburb of Halle-Neustadt. Originally planned as an independent town it is now likely that it will merge administratively with Halle. A new shopping complex serves both the old town and the new while a decision to build a town hall in the new town has been deferred.

Almost invariably state housing in Romania now takes the form of apartment blocks made largely from prefabricated concrete units. The programme is determined nationally by central planning and then disaggregated through planning in each county where there are institutes concerned with siting and design. There is however some grouping and not every county has its own organization. Building work is carried out by the local 'trust de constructii' which complements the local government planning offices. In Bucharest the development is quite conventional for large socialist cities in emphasizing the high-density complexes on the fringes, linked with industrial estates by improved transport facilities including orbital roads and railways (Figure 5.4). Space is limited but apartments have increased in size over the years and facilities have improved as well. It is now unusual for any new accommodation to be built with a level of comfort inferior to the highest grade. This means that each apartment will have a kitchen, bathroom, and balcony; there will also be no transit rooms. By comparison the second

Figure 5.4 Planning in Bucharest

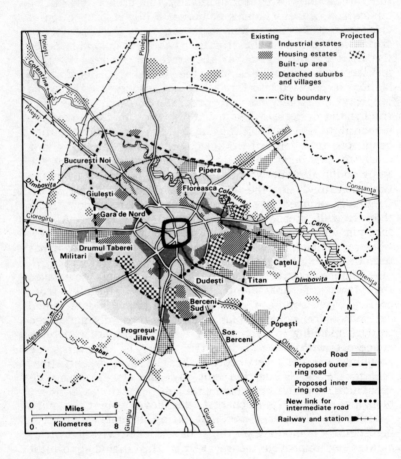

Source: D. Turnock (1974) 'Urban development in a socialist city: Bucharest', Geography 59:344-8.

level of comfort would mean only a small kitchen and bathroom and no balcony at all and the third level (now encountered only rarely) would have no water supply. The use of brick is decreasing while prefabricated concrete components are increasing. Problems may be encountered on the top floor of the block through leaking roofs and also on the first floor above ground (where the defective apartments are often allocated to gipsy families). Standardization is excessive and some fittings are of poor quality. Electricity is universally installed (though sometimes to the exclusion of gas, which is usually preferred for cooking), but there is some concern over safety through the poor quality of the plastic components and the substitution of aluminium for copper. Many tenants find themselves replacing doors or fitting double glazing, to overcome defects in building as well as to bring about improvements. As car ownership increases, the lack of garages becomes a problem. These are not normally provided in state housing projects and there appears to be no particular desire at the moment to tackle this problem since private/co-operative planning applications for garage construction in Bucharest have not been approved. For the present vehicles stay out in the open, usually covered by a waterproof sheet when the car is not being used regularly, but this is unsightly as well as unsatisfactory for security and maintenance.

There are some wider issues which arise in the housing debate and they have relevance for the quality of life in socialist cities. It may be that 'the mode in which the planners see fit for people to live and the style the people themselves prefer may be incompatible. A social problem may be taking shape which might end up looming larger than the simple question of housing itself'. (30) There is conflict between state and private goals with regard to housing and services and it is not clear how they will be resolved. The state will resort to mass media campaigns for a higher level of 'socialist consciousness' while the population will look for some compromise in the form of discrete modifications to government directives. (31) There is first the supply of housing, which is inadequate. During the 1970s in Poland for example there was a widening gap between the level of housing needs and aspirations and the opportunity to satisfy them. The present economic crisis has intensified those disproportions and removed the prospect of overcoming them. (32) Weclawowicz has undertaken a study in factorial ecology in Radom which relates housing condition to socio-occupational status. The shortage of housing and the scope for political manipulation 'accounts for the fact that access to dwellings, their standards and location, varied for different occupational and social groups and has become one of the sources of contradictions and political tension in Polish cities'. (33) Such tensions are particularly evident in Radom where even during the 1970s a substantial increase in investment in housing failed to keep pace with the increase in the number of households in the city.

Then there is the adequacy of the new apartments provided. The one- or two-roomed apartments are too small for families with several children while further irritations arise from poor soundproofing and extremes of temperature as concrete walls 'burn with heat' in the summer and remain ice cold in winter. Life is monotonous in the 'sleeping silos', all the more so when consumer services are inadequate and when the small green spaces are eaten up by increasing car-parking needs. Alienation arising out of public dissatisfaction over new apartment blocks also arises over construction defects and architecture, for new housing often appears highly monotonous. Soviet influence was very strong immediately after the war and there was a tendency to build heavy stone edifices with comparatively small windows, notably in Bulgaria, although this was not very satisfactory in towns where the climate was much warmer than in Moscow. Russian influence has now dwindled but instead architectural innovation is constrained by the dead hand of standardization evident in the construction of multi-storey blocks from prefabricated components. Variety can be introduced quite effectively through murals but some decorations have been highly inappropriate: notably the introduction of Corinthian columns in the new towns of Havirov and Poruba near Ostrava which has consigned 'Porubism' to the dictionary as an indicator of singularly ironic and pretentious urban design. Finally there is the tendency for the state to emphasize collective activities. People may be unimpressed with new community halls, playgrounds, and sports fields when so much more could be done to make life pleasanter for the individual. Some young people turn their alienation outwards in the form of vandalism while older people take a passive approach and withdraw into an exclusively private sphere.

The response by the state varies from country to country. In Romania there is a powerful propaganda effort to educate the public so that apartment blocks continue to be the norm with very little scope for individual choice. The main priority is to minimize encroachment on to agricultural land so construction of new detached houses in Romanian towns is now virtually impossible and even in the villages, where individual houses must now have at least two storeys, the blocks are making an appearance. Moreover the housing densities are extremely high, even where private or co-operative ownership of apartments occurs. However, sales of apartments by the state have been sharply curtailed since 1986, largely it would seem because of government concern over growing speculation and inequality in housing. Apartments have been purchased in areas perceived as desirable and the value of these properties has then increased by two or three times over little more than a decade as families unable to buy what they want from the state have competed for apartments previously sold to the occupants. However, in most other countries the toleration, if not encouragement, of the private/co-operative sector is leading to a measure of variety among apartments in terms of architecture, living space, and population densities. In Poland in the late 1950s some encouragement was given to co-operative housing schemes with participants paying fairly realistic prices as either owners or tenants. This led to some anomalies because rents for state accommodation were subsidized leading to inequalities in housing costs which were not related to income. Most people have to apply for the more expensive co-operative housing rather than the cheaper state accommodation. However the differences were reduced in the 1960s and then in 1972 Poles were allowed to purchase state-owned apartments on attractive terms. Together with a subsequent decision to discontinue the construction of state housing altogether this gave an important boost to private ownership both within and without the co-operative sector. (34)

In Czechoslovakia too there is scope for building by co-operatives and enterprises as well as the state and since 1966 state housing has been sold to private buyers, initially on the condition that there would be buyers for every apartment in the building but since 1978 on the basis of single apartments. In some cases attempts to shift the housing burden to private enterprise extends to provision of state credits. Such facilities are granted to workers in agriculture and forestry and to the population at large in border regions. The state is also providing some larger flats,

183

with three rooms or more, on the new estates and architectural improvements are being made through more imaginative work in the prefabrication factories, now using Soviet and Canadian equipment and greater activity in the design centres (such as Stavoprojekt in Brno). Moreover the encouragement of individual and co-operative enterprise provides some scope for a diversification of new urban landscapes away from the apartment block. On privately owned land close to the urban fringe largely unplanned settlements of rural-type housing may be tolerated by the authorities; for example, Kozarski Bok and Trnje near Zagreb and Nowy Dwor near Warsaw. In environmentally sensitive areas of Budapest small-scale housing projects are being planned in the state sector, for example in the Csatarka-Torokvesz area of Buda where the hilly terrain is conducive to the building of traditional houses with only occasional tall apartment blocks.

But the private sector is also very active and there are rising residential land values as 'the hillsides of Buda are at present under the ferocious onslaught of the small private co-operative house' so much so that conflict is arising with the ideal of landscape conservation and the government's desire to limit speculation. (35) In Hungary there are substantial inducements to private builders such as low-cost rents for building plots and long-term loans to cover building costs (with repayment obligations automatically moderated as the number of children increases). (36) There is also encouragement for small building co-operatives so that two or more families may obtain a plot and build the appropriate number of uniform attached houses in it, possibly making some use of prefabricated units from the concrete factories which are now trying to gear production to the family house market. The state thus passes some of the responsibility for house-building on to the private sector but in a manner conducive to better environmental conditions. In Hungary only a fifth of the new housing is now financed exclusively by the state and the level of private ownership among the entire housing stock is approximately three-quarters.

CONCLUSION

One might well ponder the reasons for the interest of some western geographers in demonstrating that socialist cities are distinctive. It may arise from the stage reached in philosophical debate as criticisms of positivism create a disposition to reject the global application of western models. It may be that caveats issued by the model builders themselves have been taken very much to heart. With regard to urbanization Berry stresses the 'fundamentally different processes that have arisen out of differences in culture and time'. (37) And possibly communist claims to ideological distinction have stimulated a search for a specific urban geography which runs a risk claiming that minor differences rather than underlying similarities with western cities are 'fundamental'. Zelinsky's hypothesis of mobility transition, postulating increased levels of migration and commuting as societies become more advanced, is rejected by Fuchs and Demko in the case of Eastern Europe, claiming that 'fundamental differences become apparent when we compare the evolution of mass commuting in Eastern Europe with that in North America'. (38) Certain aberrations are mentioned which have blunted the admitted attracting force of pre-socialist cities: the concentration of early post-war industrial development in a small number of centres, a measure of involuntary commuting by peasants arising from the urban housing shortage aggravated by discrimination in the allocation of apartments which results in the anomaly of out-commuting by professional people to jobs in the countryside. Yet such aberrations fly in the face of socialist ethics and hardly amount to ideologically induced arrangements which justify a denial that Eastern Europe is 'fated' to follow western experience because of the 'determination of socialist planners to develop urban forms distinctive to socialism'. (39) While it is the geographer's task to identify and interpret differences across the world there seems little to be gained from the exaggeration of such differences to the point where fundamental conceptual similarities (such as the logic of central place hierarchy) are denied or simply ignored.

NOTES

1.   Hansen (1978); Musil (1981); Sarfalvi (1975). See also L.A. Kosinski (1974) 'Urbanization in East Central Europe after World War Two', East European Quarterly 8: 129-53; N.J.G. Pounds (1971) 'The urbanisation of East Central and Southeast Europe', in Hoffman (ed.) 54-82.

2.   V. Cucu (1972) 'Economic and demographic

premises in the urbanization of the SRR', Revue roumaine: geographie 16:103-9.

3.   K. Dziewonski et al. (eds) (1981) 'Population and settlement structure: analytical and policy approaches', Geographia Polonica 44: 5-259; K. Dziewonski et al. (1984) 'Settlement systems in centrally-planned economies: settlement in old lands and high density regions', in Bourne et al. (eds) 359-411; P. Ronnas (1982) 'Centrally-planned urbanization: the case of Romania', Geografiska Annaler 64B: 143-51; A. Simic (1973) 'Urbanization and modernization in Yugoslavia: adaptive and maladaptive aspects of traditional culture', in M. Kenny and D. Kertzer (eds) Urban Life in Mediterranean Europe: Anthropological Perspectives, Urbana': University of Illinois Press, 76-108; G.M. Zovanyi (1986) 'Structural change in the system of urban places: the twentieth century evolution of Hungary's settlement network', Regional Studies 20: 47-71.

4.   R.J. Fuchs and G.J. Demko (1977) 'Commuting and urbanization in the socialist countries of Europe', Bulletin of the Association of Comparative Economic Studies 19: 21-38; R.J. Fuchs and G.J. Demko (1977) 'Commuting in the USSR and Eastern Europe: causes, characteristics, consequences', East European Quarterly 11: 463-75; T. Gilberg (1980) 'Peasant workers in Romania', in W.P. Avery et al. (eds) Rural Change and Public Policy, Oxford: Pergamon, 179-94; I. Ianos and M. Olaru (1980) 'Contribution to the study of daily commutation in Caras-Severin county', Revue roumaine: geographie 24: 161-6.

5.   G. Konrad and I. Szelenyi (1974) 'Social conflicts of underurbanization', in A. Brown et al. (eds) I: 206-26.

6.   W.H. Berentsen (1982) 'Changing settlement patterns in the GDR 1945-1976', Geoforum 13: 327-37; A. von Kanal (1981) 'Development of settlement system in rural regions of the GDR', Geographia Polonica 44: 123-31; B. Kronert (1981) 'City hinterland regions of large cities and medium sized towns in the GDR', Geographia Polonica 44: 139-50; H. Ludeman and J. Heinzmann (1978) 'On the settlement system of the GDR', in Hansen (ed.) 121-43; C. Rosenkrauz (1975) 'Commune categories: a method of determination of conditions and tendencies in the Halle-Leipzig agglomeration', Geographia Polonica 30: 125-32; H. Schmidt (1975) 'Developing trends in the industrial structure of urban agglomerations in the GDR', Geographia Polonica 30: 61-8.

7.   J.B. Borchert (1975) 'Economic development and

population distribution in Albania', Geoforum 6: 177-86; M. Macka et al. (1981) 'Selected aspects of the internal organisation of a settlement system', Geographia Polonica 44: 65-88; A. Zagozdzon (1968) 'Problem of local settlement complex', Geographia Polonica 14: 345-52; A. Zagozdzon (1972) 'Semi-urbanization with special reference to the problem of the settlement complex', Geographia Polonica 24: 57-69; A. Zagozdzon (1973) 'Problems of development of a settlement network in a region under industrialization', Geographia Polonica 27: 159-74.

8. Kosinski (1977); Kostanick (1977); Sarfalvi (1971). See also P.A. Compton (1969) 'Internal migration and population change in Hungary between 1959 and 1965', Transactions Institute of British Geographers 49: 111-30; P.A. Compton (1972) 'Internal migration in Hungary between 1960 and 1968', Tijdschrift voor Economische en Sociale Geografie 63:25-38; A. Gawryszewski (1977) 'The role of permanent migration and commuting in urban growth', Geographia Polonica 37: 47-59; G.W. Hoffman (1973) 'Currents in Yugoslavia: migration and social change', Problems of Communism 22(6); 16-31; L. Kosinski (1970) 'Internal migration of population in Poland 1961-1965', Geographia Polonica 18:75-84.

9. French and Hamilton (1979). See also P.A. Lentz (1975) 'A model of residential structure in a socialist city: a case study of Warsaw', Geographia Polonica 31:65-97; J. Musil and Z. Rysavy (1983) 'Urban and regional growth processes under capitalism and socialism: a case study from Czechoslovakia', International Journal of Urban and Regional Research 7: 495-527; D. Turnock (1974) 'Urban development in a socialist city: Bucharest', Geography 59: 344-8.

10. P. Beluszky (1971) 'Central places developing in County Szabolcs-Szatmar', in Sarfalvi (ed.) 165-81.

11. F.W. Carter (1979) 'Prague and Sofia: an analysis of their changing city structure', in French and Hamilton (eds) 425-49 (quote p. 454).

12. I. Chudzynska (1981) 'Spatial structure in retail trade in Warsaw', Geographia Polonica 44: 201-10; I. Chudzynska (1981) 'Locational specialization of retail trade functions in Warsaw', Environment and Planning 13A: 929-49; J. Grocholska (1975) 'Possibilities of determining the factors that affect urban land use: a case study of Warsaw', Geographia Polonica 31: 53-63; K. Polarczyk (1976) 'Distribution of service centres within large urban areas: a

market accessibility model', Geographia Polonica 33: 143-55; J. Rakowicz-Grocholska (1970) 'Investigation of urban land use in Polish geographical studies', Geographia Polonica 18: 85-92.

13.   J. Dzieiuchowicz (1981) 'Spatial distribution of intraurban commuting to work: a case study of the city of Lodz', Geographia Polonica 44: 211-23; Z. Taylor (1976) 'Accessibility of urban transport systems: the case of Poznan city', Geographia Polonica 33: 121-41.

14.   K. Grime and G. Weclawowicz (1981) 'Warsaw', in M. Pacione (ed.) Urban Problems and Planning in the Developed World, London: Croom Helm, 258-91; G. Prawelska-Skrzypek (1988) 'Social differentiation in old central city neighbourhoods in Poland', Area 20: 221-32; J. Regulaska (1987) 'Urban development under socialism: the Polish experience', Urban Geography 8(4): 321-39; G. Weclawowicz (1977) 'The structure of socioeconomic space in Warsaw 1931 and 1970', Geographia Polonica 37: 201-24; G. Weclawowicz (1979) 'The structure of socioeconomic space in Warsaw 1931 and 1970: a study in factorial ecology', in French and Hamilton (eds) 387-423 (quote p. 422).

15.   A. Burnett (1985) 'Inequalities under state socialism: what is the evidence on environment differentiation at the neighbourhood scale?', in L. Collins (ed.) Anglo-Bulgarian Symposium 1982, London: University of London School of Slavonic and East European Studies, 2: 38-51; R.J. Fuchs and G.J. Demko (1979) 'Geographic equality under socialism', Annals Association of American Geographers 69: 304-18; I. Szeleny (1983) Urban inequalities under state socialism, Oxford: Oxford University Press.

16.   G. Church (1979) 'Bucharest: revolution in the townscape art', in French and Hamilton (eds) 493-506 (quote p. 502).

17.   French and Hamilton (eds) 6.

18.   ibid. 200.

19.   ibid. 178.

20.   ibid. 453.

21.   G. Weclawowicz (1985) 'The socio-spatial structure of Radom city in 1978', Geographia Polonica 51: 85-98 (quote p. 85).

22.   French and Hamilton (1979) 184.

23.   ibid. 286.

24.   Rugg (1985) 364-6.

25.   A.H. Dawson (1971) 'Warsaw: an example of city

structure in free-market and planned socialist environments', Tijdschrift voor Economische en Sociale Geografie 62: 104-13 (quote p. 112).

26.   P.A. Compton (1979) 'Planning and spatial change in Budapest', in French and Hamilton (eds) 460-91 (quote pp. 490-1).

27.   Bahro (1978) 7-8.

28.   A.M. Weclawowicz (1985) 'Some aspects of crisis housing in Poland', Geographia Polonica 51:99-112.

29.   D. Turnock (1989) 'Romania', in Sillince (ed.).

30.   M. MacQueen (1980) 'The housing problem in the GDR', Radio Free Europe Background Report, 189 (quote p. 1).

31.   J.C. Fisher (1962) 'Planning the city of socialist man', Journal American Institute of Planners 28: 251-65; Kansky (1976).

32.   G. Weclawowicz (1985) 'The socio-spatial structure of Radom city in 1978', Geographia Polonica 51:86.

33.   ibid. 85-6.

34.   M. Ball and M. Harloc (1974) 'Housing policy in a socialist country: the case of Poland', Centre for Environmental Studies Research Paper; W. Malicka (1979) 'Housing estates and town communities in post-war Poland', International Journal of Urban and Regional Research 3: 209-19; W. Sieminski (1979) 'The social goals of residential communities in Poland', International Journal of Urban and Regional Research 3: 220-7.

35.   P.A. Compton (1979) 'Planning and spatial change in Budapest', in French and Hamilton (eds) 483; J.A.A. Sillince (1985) 'The housing market of the Budapest urban region, 1948-1983', Urban Studies 22: 141-9.

36.   J. Hagedus and I. Tosics (1983) 'Housing classes and housing policy: some changes in the Budapest housing market', International Journal of Urban and Regional Research 7: 467-94; J. Rowland (1986) 'Private self-build begins to take hold', Town and Country Planning 55: 318-20; J.A.A. Sillince (1985) 'Housing as social problem versus housing as historical problem; the case of Hungary', Environment and Planning 3C: 299-318.

37.   B.J.L. Berry (1973) The Human Consequences of Urbanization, New York: St Martin's Press, xii.

38.   R.J. Fuchs and G.J. Demko (1978) 'The post-war mobility transition in Eastern Europe', Geographical Review 68: 171-82 (quote p. 175).

39.   ibid. 180.

Chapter Six

PLANNING THE GROWTH OF LARGE TOWNS

A very obvious feature of the urbanization process has been the increasing prominence of larger towns with a population exceeding 50,000. The number of such towns rose from 150 in 1965 to 257 in 1980 (Table 6.1). Towns with more than 100,000 inhabitants increased from 67 to 105 during the same period. The larger towns seem to be growing most rapidly at the present time for the urban system appears to have achieved some stability after the early post-war years of administrative reorganization and new town development in some countries. Towns with more than 50,000 inhabitants accounted for 47.8 per cent of the urban population of Poland in 1950 rising to 53.8 in 1970 and 62.8 per cent in 1980. This is because of the significant migrational element in the growth patterns of the larger cities (Table 6.2).

All Polish towns except eleven (all of them components of urban complexes) experienced natural increase during the 1970s but 232 have experienced a net migration loss and 84 per cent of these cases involved small towns with a population below 10,000 (61 per cent for towns with fewer than 5,000 inhabitants). By contrast only a few large cities suffered net out-migration, mostly towns in the Jelenia Gora and Walbrzych regions, where obsolete industries are declining, or in the Katowice area, where decongestion policies are in force. Taking all towns with a population exceeding 20,000 net in-migration contributed more to growth in the early 1970s than did natural increase whereas the reverse was true in the lower categories. The role of migration has been quite outstanding in some individual cases and Dobrowolska's survey of Stalowa Wola and Tarnobrzeg in the Rzeszow region shows that nearly 45,000

Table 6.1 Distribution of cities by size groups 1965 and 1980

Country	1965				1980			
	A	B	C	D	A	B	C	D
Albania	6	0	1	0	4	5	1	0
Bulgaria	22	6	2	1	31	12	6	1
Czechoslovakia	34	13	4	1	58	24	6	1
GDR	77	13	9	2	81	20	12	3
Hungary	34	9	4	1	41	14	7	1
Poland	73	23	19	3	99	39	30	5
Romania	37	6	12	1	33	22	18	1
Yugoslavia	47	11	6	1	55	14	11	2
Eastern Europe	331	83	57	10	403	152	91	14

Source: Statistical yearbooks.

Notes: 1980 figures for Bulgaria and Yugoslavia are estimates.
A 20,000–50,000 B 50,000–100,000 C 100,000–500,000 D Above 500,000

Table 6.2 Urban growth in Poland 1950–80 by size groups

Size groups	1950			1980			Growth indexes 1970–5	
	A	B	C	A	B	C	D	E
Below 5,000	393	10.9	1.01	270	4.9	0.99	4.10	1.69
5,000–19,999	217	20.6	1.90	351	17.9	3.66	1.22	0.88
20,000–99,999	80	26.4	2.44	147	31.3	6.33	0.75	0.82
100,000 and above	16	42.1	3.89	35	45.9	9.31	0.65	1.09
Poland	706	100.0	9.24	803	100.0	20.29	0.82	1.00

Sources: M. Jerczynski et al. (1981) 'Development of the national system of cities as related to migration', Geographia Polonica 44: 97–109; Statistical yearbooks.

Notes:
A Number of towns
B Percentage share of urban population
C Total population (millions)
D Growth by natural increase in relation to growth by migration
E Share of total urban population related to share of growth 1970–5

people settled during the 1960s, including nearly 9,000 from outside the region. (1)

NATIONAL CAPITALS

The most dynamic cities tend to be the national capitals which offer a wide range of job opportunities and a relatively stimulating cultural life. (2) However the primacy of the capitals varies very considerably across Eastern Europe. Table 6.3 reveals the general picture. Capital cities grew by 56.5 per cent between 1950 and 1980, a slightly lower rate than for the ten largest provincial cities (84.2 per cent) and the urban population as a whole (99.4). But a higher proportion of the total population now inhabits the capital cities (7.5 per cent in 1980) and the dominance over the second city is undiminished with a 3:1 ratio maintained in Eastern Europe as a whole. There are however remarkable contrasts between individual countries according to historical circumstances. Some were imperial capitals before 1918 and accommodate a high proportion of the total urban population of the countries concerned.

In the case of Budapest the dominance of the city was increased by the break-up of imperial Hungary after the First World War; Dienes has made dramatic reference to the distortion introduced by frontier changes which placed the major provincial cities of imperial Hungary within boundaries of the successor states:

> Especially on the Great Plain the regions between these centres and Budapest had been economic troughs which were siphoned clean of their most enterprising elements and where commerce and industry could find no sustenance. Within the new boundaries therefore and especially on the south and east Budapest has been surrounded by economically weak and underdeveloped regions linked to the capital in almost complete dependency. (3)

By contrast in the GDR the dominance of the capital city is moderated by the division of Berlin which leaves the majority of the population outside the socialist state. Federal arrangements also moderate the demographic weight of the capital city even when growth has been sustained throughout modern times. In Czechoslovakia the

193

Table 6.3 Population of capital cities in Eastern Europe 1950-80

Country/capital	Year	Population (thousands)			Population (millions)		Capital (per cent) related to			
		A	B	C	D	E	B	C	D	E
Albania (Tirana)	1950	40^e	29	143	0.25	1.21	137.9	28.0	16.0	3.3
	1960	130	41	227	0.51	1.63	317.1	57.3	25.5	8.0
	1967	169	53	332	0.80	2.02	318.9	50.9	21.1	8.4
	1978	198	62	402	0.90	2.59	319.3	49.3	22.0	7.6
Bulgaria (Sofia)	1947	437	127	505	2.00	7.15	344.1	86.5	21.8	6.1
	1959	671	171	802	3.01	7.87	392.4	83.7	22.3	8.5
	1970	868	247	1,236	4.51	8.47	351.4	70.2	19.2	10.2
	1980	1,057	350	1,627	5.51	8.88	302.0	65.0	19.2	11.9
Czechoslovakia (Prague)	1950	933	285	1,129	6.35	12.51	327.4	82.6	14.7	7.5
	1960	1,005	314	1,323	7.89	13.74	320.1	76.0	12.7	7.3
	1970	1,078	335	1,572	8.94	14.34	321.8	68.6	12.1	7.5
	1980	1,193	375	1,914	10.16	15.28	318.1	62.3	11.7	7.8
GDR (East Berlin)	1950	1,090	618	2,633	13.04	18.50	176.4	41.4	8.4	5.9
	1963	1,072	590	2,600	12.37	17.19	181.7	41.2	8.7	6.2
	1971	1,085	585	2,658	12.59	17.06	185.5	40.8	8.6	6.4
	1981	1,146	563	2,735	12.75	16.74	203.6	41.9	9.0	6.8

Hungary (Budapest)	1950	1,600[e]	128	693	3.55	9.31	1,250.0	230.9	45.1	17.2
	1960	1,807	144	843	4.34	9.98	1,254.9	214.4	41.6	18.1
	1970	1,940	173	1,028	4.99	10.31	1,121.4	188.7	38.9	18.8
	1980	2,061	210	1,354	5.70	10.71	981.4	152.2	36.2	19.2
Poland (Warsaw)	1950	804	620	2,701	9.61	24.98	129.7	29.8	8.4	3.2
	1960	1,139	710	3,462	14.40	29.78	160.4	32.9	7.9	3.8
	1970	1,316	763	4,076	17.09	32.67	172.5	32.3	7.7	4.0
	1981	1,617	843	4,871	20.29	35.73	191.8	33.2	8.0	4.5
Romania (Bucharest)	1948	1,042	118	914	3.71	15.89	883.1	114.0	28.1	6.6
	1958	1,279	161	1,212	5.99	18.06	794.4	105.5	21.4	7.1
	1970	1,475	203	1,806	8.34	20.25	726.6	81.7	17.7	7.3
	1980	1,861	305	2,652	11.01	22.20	610.2	70.2	16.9	8.4
Yugoslavia (Belgrade)	1950	389	291	1,054	3.27	15.77	133.7	36.9	11.9	2.5
	1960	588	427	1,402	5.24	18.55	137.7	41.9	11.2	3.2
	1970	755	562	2,023	7.38[e]	20.53	134.3	37.3	10.2	3.7
	1980	911	659	2,441	10.00[e]	22.30	138.2	37.3	9.1	4.1
Eastern Europe	1950	6,335	2,216	9,772	41.78	105.32	285.9	64.8	15.2	6.0
	1960	7,691	2,558	11,871	53.75	116.80	300.7	64.8	14.3	6.6
	1970	8,686	2,921	14,731	64.64	125.65	297.4	59.0	13.4	6.9
	1980	10,044	3,367	17,996	83.32	134.43	298.3	55.8	12.1	7.5

Source: Statistical yearbooks.

Notes: A Capital city
B Largest provincial city
C Top ten provincial cities
D Total urban population
E Total population
e Estimate

Figure 6.1 Urban growth and the development of industrial centres in Romania

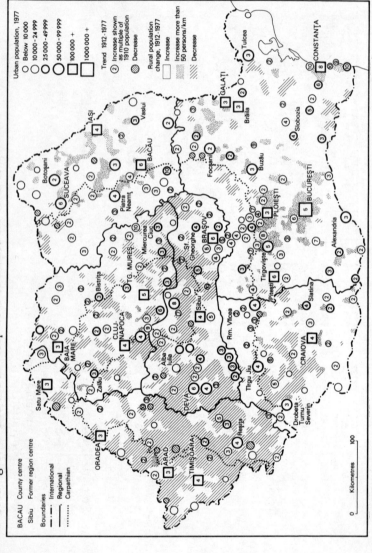

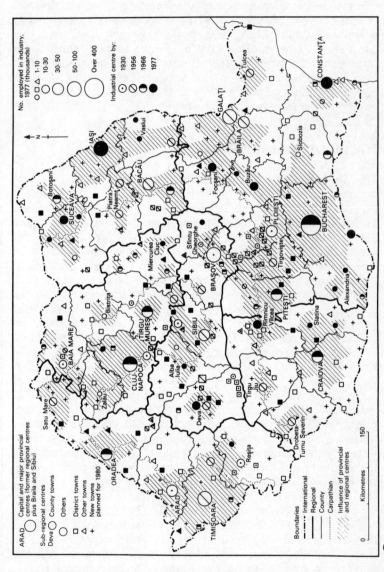

Source: Census.

rapid growth of Bratislava since 1968, as capital of Slovakia, has served to reduce Prague's share of the total population of the country. (4) In Yugoslavia Belgrade, as the federal capital, is complemented by the capitals of the individual republics with Zagreb (capital of Croatia) particularly prominent while in Albania, where government is highly centralized, the capital city of Tirana is not many times larger than Shkoder because it was a small provincial town in the Ottoman Empire until it was selected as the Albanian capital, whereas Shkoder was an important administrative centre in the imperial period. (5)

THE INDUSTRIAL FUNCTION

The importance of large cities rests on a wide range of services provided for an extensive hinterland. Special shops are comparatively rare by western standards but there are important cultural facilities such as theatres and universities, and important opportunities for tourism and recreation. Prague's Fucik Park for rest and culture is dominated by a congress hall which is used for major conferences (including Communist party congresses), exhibitions, and other functions. The cinema, concert hall, and planetarium provide for a range of cultural activities and the park also includes various sports facilities in addition to botanical and zoological gardens. Administration is also important, relating to regional planning, the Communist party organization, and trade unions. But manufacturing remains fundamental, and the growth of large cities has gone hand in hand with the development of industrial centres, as the case of Romania demonstrates (Figure 6.1). Large cities are necessary to accommodate the labour and infrastructure required by major industries. (6) In Ostrava the Vitkovice and Klement Gottwald enterprises (concerned with metallurgical and foundry work) employ more than 75,000 workers in engineering while jobs at both the Skoda works in Plzen and the CKD works of Dukla and Vysocany in Prague exceed 40,000. The main engineering works in Brno (Zbrojovka, ZKL, and Kralovopolska Strojirna) together employ more than 30,000 while there are more than 12,000 jobs in oil refining, chemistry, and petrochemistry at Slovnaft and Juraj Dimitrov in Bratislava. Altogether the six main cities in Czechoslovakia (Plzen and Prague in Bohemia, Brno and Ostrava in Moravia, and

Bratislava and Kosice in Slovakia) with 15.5 per cent of the population account for 20.5 per cent of industrial output by value. Two-thirds of the GDR's Associations of Nationally Owned Enterprises (Vereinigung Volkseigener Betriebe: VVB) have their head offices in the five largest cities: East Berlin, Leipzig, Karl Marx Stadt, Dresden, and Halle.

Even where non-industrial functions have particular importance the emphasis on manufacturing is maintained. Thus Prague as a socialist city must retain some importance for production and not become the exclusive preserve of white-collar workers. Manufacturing industry may be relocated if it is obviously harmful to a sensitive environment but there can be no objection to electronics and precision engineering and extensive industrial areas continue to be included in plans for new housing complexes. In the city of Lodz there has been some redevelopment (despite only moderate bomb damage) and some removal of industry from the main streets but the government remains committed to an industrial strategy for the city and the density of industrial employment in the city centre is now higher than it was before the war, while new industrial estates have appeared on the outskirts. (7) The greatest changes have been in industrial structure through curtailment of expansion of the textile industry to allow for its expansion in parts of rural Poland.

The party may well have a political motive for the emphasis on industry, having regard to the significance of a large working-class population to counter potential opposition from the intelligentsia. Certain cities seem to have a capacity for opposition arising from their cultural and social history. Thus Jena in the GDR has a history of friction and confrontation with the holders of political power. (8) The city played an important role in the workers' revolt of 1953 and displayed a negative reaction to the Warsaw Pact invasion of Czechoslovakia in 1968. Most recently there has been support for peace groups which have emerged outside the institutional confines of the Thuringian Church as well as the SED, which sponsors its own peace council. Such activity is the work of a sophisticated urban community, offering illegal political opposition over issues which the state can hardly condemn in the light of its own premise that peace and socialism are inherently linked. Another example would be Krakow, where the Polish party sought to increase its influence by constructing a new town and metallurgical complex (Nova Huta) on the eastern edge.

THE LANDSCAPE OF THE LARGE CITY

Even in the large cities the central business district is poorly developed by western standards. The limited amount of private ownership of land in city centres and the strict rent controls result in the absence of a land value gradient from centre to periphery. Of course there is a definite logic in placing appropriate functions in the centre of the city where the main transport routes intersect but the limited spending power of the average family and the absence of competition in retailing, banking, and insurance means that the central area need not be devoted exclusively to such service functions. Various political and cultural functions are prominent, for example higher education/research institutes, museums, libraries, trade union organizations; residential functions are much in evidence with multi-storey buildings providing space for shops at the ground level and apartments above. Hence the need for a considerable number of food shops in the central area as well as special shops catering for the occasional extravagance. The state's interest in distributing approved political literature gives bookshops a prominence which is not matched in the western city in terms of the floor-space available and the accessible sites which are selected. The central area is very likely to offer scope for the tourist by way of routine facilities for the traveller and a legacy of historic buildings dating back to medieval (and in some cases classical) times. Only rarely were town centres reconstructed before the socialist period: one unusual case in Szeged in Hungary, a city destroyed by flooding in 1879. The centre was filled in and rebuilt at a higher level with a system of boulevards. The historical legacy is usually very strong and even in towns which have experienced very rapid recent growth the former agrarian characteristics may still be very apparent as at Kecskemet. (9)

Outstanding examples like Krakow and Prague present serious conservational problems which are examined elsewhere whereas certain other cities which experienced heavy war damage followed by redevelopment in the 1950s and 1960s may retain only a few historic buildings in a core given over largely to new housing, modern central functions, and open space. However in their own way the rebuilt centres display some remarkable monumental architecture. Usually ad hoc redevelopments are in evidence, perhaps relating to sites affected by bombings, earthquakes, or floods, and these tend to complicate comprehensive renewal at a later stage. The problems are increased by the lack of finance for the proper maintenance of the better houses. Although owner-occupiers are obliged to accept lodgers where their space exceeds the norms laid down by the state, the income is too small to cover the cost of repairs and landlords have been known to offer property free to the state in order to escape from a burdensome responsibility. Perhaps the most radical changes have occurred where space has been taken to extend the city centre. The outstanding example is Katowice, where a new centre has been built since the mid-1960s extending for almost a kilometre north from the old centre with its contrasting architecture and functions.

Adjacent to the historic core are mixed zones of housing and industry relating to the railway age. These are most extensive in cities which experienced substantial industrialization before the Second World War: the major cities of the German Empire (to a much lesser extent the Habsburg Empire) and the Balkan capitals. Industry and commerce remains prominent, usually with consolidation of the business structure to create a smaller number of enterprises, and housing quality ranges from the well-appointed properties of the former bourgeoisie (frequently used today as embassies, hospitals, institutes, publishing houses, etc.) to poor rural-type houses often lacking piped water, proper sanitation, and surfaced roads. Some comprehensive development has occurred but for the most part the resources for housing are too limited to allow for widespread demolition of the sub-standard accommodation beyond the building of apartment blocks along the main highways. The entire urban landscape is being transformed, but it is on the periphery where the construction of new neighbourhoods has been most marked. All the main towns have experienced the building of large new housing complexes dominated by five-storey apartment blocks, embodying the ideal of communal living with each unit having its dormitories, dining-rooms, kindergartens, schools, shops, dispensaries, and local industries. However the communal living concept has not worked well in practice and while there may be some local employment available it is usual to find workers travelling to all parts of the city for work. Co-operation usually survives only in the administration of the block which is typically one of a large number of such buildings grouped together on state-owned

Figure 6.2 Housing in Cluj–Napoca, Romania

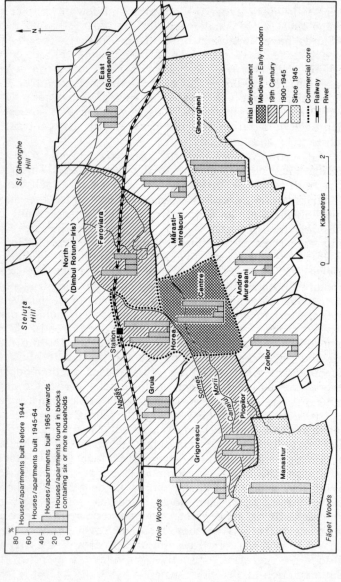

Source: Field-work.

land. There is inevitably a certain monotony in the landscape although there is usually greater variety in points of detail in the more recent blocks than in the heavily stereotyped buildings of the Stalin period. But perhaps the most outstanding feature is the overcrowding through the close spacing of blocks and the over-representation of two-room apartments which families with several children frequently have to use. Planners freely admit that the situation cannot continue indefinitely and that the new blocks may only have a short life. Yet for the foreseeable future government investment priorities allow no real alternative and for most young couples the apartment in town is the height of their expectations.

Case studies from Romania: Cluj-Napoca and Iasi

These examples are useful in demonstrating first that modern high-rise building has encroached into all the neighbourhoods of large cities (although the persistence of traditional housing continues to give some quarters a distinctive character and appeal) and second that the physical environment can exert a powerful influence through variations in the suitability of land for building. It is evident from Figure 6.2 that in the city of Cluj-Napoca the proportion of households resident in accommodation built since 1945 (overwhelmingly in blocks) is now substantial in all districts. (10) There are individual apartment blocks on restricted opportunity sites (common in the central area) and larger complexes which call for the demolition of existing housing (usually single-storey rural-type detached houses with substantial gardens). Of course apartment blocks are by no means restricted to the post-war period, although the very tall blocks with up to ten storeys and the very long buildings (around 100 metres) are characteristic of this period alone. If blocks are regarded as buildings containing six or more apartments then it would appear that they account for proportions in excess of the share of post-1945 buildings by the approximate margins indicated on the map. As for other variations between districts these can be discussed only impressionistically because of the lack of reliable evidence. However it would appear that areas consisting of substantial houses built in the nineteenth and early twentieth centuries and situated in a pleasant environment reasonably close to the centre are regarded as particularly desirable, so that higher-income families are able to compete effectively for rented or owner-occupied accommodation.

Generally speaking all pre-1945 accommodation seems to be held in high regard because the high densities characteristic of new housing complexes are avoided. The only difficulties are large rooms which give rise to high heating costs and make it difficult to rent small areas of surplus space. It is not unusual to find vacant apartments in some of the large peripheral housing complexes because the more affluent families will try and find older properties in the lower part of the city. Such tendencies may be exaggerated where there are perceived variations in shopping, transport, and amenities. In other words while there may be only modest differences in living space per person and in possession of basic household equipment, variations tend to be more marked in terms of proportions of households where the head of the family is an intellectual, with higher education.

In allocating land for the construction of apartment blocks attention must be given to micro-climate (wind direction and aspect) to ensure reasonable shelter and direct sunlight. Terrain is an important factor where slopes exceed 5 per cent. On slopes of between 5 and 8 per cent long buildings need to be placed perpendicular to the slope and on even steeper slopes there are substantial increases in building costs, although greater dispersal of building to avoid broken ground and ravines may result in higher servicing costs. But planning must always take careful note of abrupt slopes running down from the edge of low plateaux. It is quite common for the more prominent blocks to be situated close to the edge in order to enhance the impressiveness of the townscape, as at Cluj (Gheorgheni), Galati (Tiglina), and Pitesti (Calea Craiovei) where ten-storey blocks on the edge of a broken plateau effectively mark the urban limit (though the blocks on the edge of the Danube bluffs at Tiglina have been considered somewhat monotonous). Blocks of four or five storeys (without lifts) are usual under normal geotechnical conditions. Where there are difficulties there can be a reduction in height or there can be more elaborate foundations and a greater volume of construction, perhaps compensated for by an increase in the number of apartments (and reduction in average size). Since the very serious earthquake of 1977 it has been necessary to give greater attention to the seismic factor with appropriate modification to building systems and to the height and disposition of blocks particularly in Vrancea and other areas close to the Carpathian 'curve' including Bucharest.

As regards the physical basis for house-building particular interest attaches to the studies in urban geomorphology carried out at Iasi University during the 1950s. Many of the leading towns of Moldavia were covered but the most detailed attention was given to the city of Iasi where severe physical constraints required the planning of a more compact settlement (Figure 6.3). The studies in physical geography were made in conjunction with a general plan of 1945-6 and a more detailed programme drawn up in 1954. Iasi is, of course, an historic city, being the capital of Moldavia from the 1550s until Bucharest became capital of the unified principalities of Moldavia and Wallachia in 1862. Its national importance arose from its central position in Moldavia (prior to the loss of Bessarabia to the Russian Empire in 1812) and its commercial role related to the trade routes running south from Poland to the lower Danube. But there is also a local marketing function enhanced by the contact between the Moldavian Plateau and the lower ground of the Prut valley and southern Moldavia. The chosen site was the northern side of the asymmetrical Bahlui valley (15 kilometres from the Prut) where there was a well-developed terrace system above the flood plain. The terraces provided not only a dry site sheltered from north winds but also one endowed with springs (for sources at Ciric and Sapte Oameni sustained the first organized water supply at the Golia monastery in the early eighteenth century). But above all the terraces are relatively untroubled by the land-sliding which affects much of the hill and plateau country of northern Moldavia.

Figure 6.3 shows first the alluvial plain of the Bahlui and the dejection cone of the Nicolina. The whole of this area, up to 2 kilometres wide, was inundated during the catastrophic floods of 1932, although moderate flooding (arising through rapid run-off from impermeable ground) is frequent and occurred fifteen times during the period 1921-53. This is not an area where housing can be encouraged although with the regulation of the Bahlui and the construction of water storages on tributary streams in the vicinity of Iasi (Ciric and Chirita) and further afield (the Belcesti was completed in 1974) it is much more feasible to contemplate industrial development, adjacent to the main

Figure 6.3 Suitability of land for building in Iasi, Romania

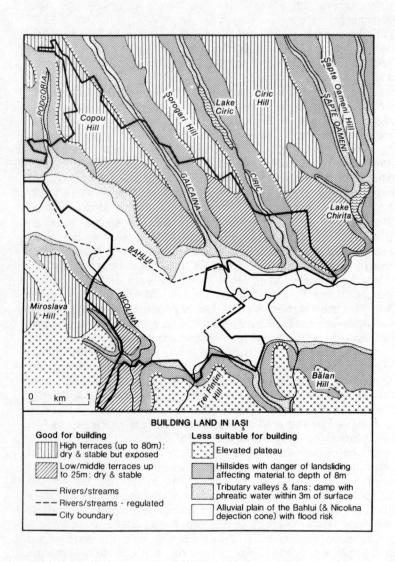

Sources: Romanian researches, especially C. Martiniuc and C. Bacauanu (1959) 'Harta geomorfologica a orasului Iasi', Analele stiintifice ale Univ. A.I. Cuza din Iasi II 5: 183-90.

road and railway. Immediately above the floodplain proper is a narrow strip of damp ground comprising the bottom lands of the tributary valleys and the fans developed by such streams in the main valley of the Bahlui. Such land is rich in underground water with a hydrostatic layer at a depth of 1-3 metres (compared with 0-2.5m on the floodplain) and even brick houses built in clusters are likely to suffer from problems of damp. However the terraces are much more suitable for building and the system extends from the lowest terrace at 10m (Frumoasa-Nicolina) to 25m (Palat-Tatarasi), 80m (Ciric/Copou/Galata), and finally 140-170m (Aroneanu/ Breazu/Miroslava). The terrace system is restricted south of the Bahlui (with only a 10m terrace at the Nicolina-Bahlui confluence) but well developed on the northern sides especially between the Galcaina and Copou streams where the medieval town developed, protected from attack from the south by the Bahlui floodplain and the outpost forts of Cetatuia and Galata on the south side of the valley.

The first stage of development occupied the lower and middle terraces up to 25m. These terraces are of Upper Pleistocene age and have a covering of loess to a depth of 5-8m. Later growth, occurring between the eighteenth and twentieth centuries, occupied the upper terraces on the same Copou ridge. These are of Middle Pleistocene origin and have a thicker loess layer (10-15m). The ground was sufficiently elevated for wind power to be exploited, as recalled in street names like Str Moara de Vint (referring to one of the nineteenth-century mills). Development also affected the lower terrace across the Galcaina river in Tatarasi, at the southern extremity of Dealul Sorogarilor. The main thoroughfare of the city ran south-east to north-west (Str Stefan cel Mare, Str Lapusneanu, Str 23 August, and Aleea Ghica Voda) following the sequence of terraces on the Copou ridge, beginning at the site of the old fortress (replaced by the Culture Palace, which was finished in 1916) and proceeding past the main square (Piata Unirii) and the University to reach the suburb of Copou. The rest of the land consists of steeply sloping, hilly ground which is generally unstable. Landslips occur naturally on account of the alteration of permeable and impermeable strata and they are known even in areas of long-standing woodland. But the problem has been accentuated by the clearance of land for agriculture and some slides now have great depth (tens of metres) on marly ground. Conditions vary locally according to the underground water circulation and the geological layering (clay, sand, and marl). In places the relief develops a certain plasticity and serious sliding results, as in 1942 when several hundred houses were destroyed in Ticau. There are several prominent scarps and edges where protective works are concentrated and the relief culminates in chains of low hills: those of Ciric, Copou, Sarogari, and Sapte Oameni to the north and Balan, Caprita, Cetatuia, Miroslava, and Trei Fintina to the south.

The development plan must, therefore, concentrate attention on the terraces, maximizing opportunities at Tatarasi and extending growth further east to the Ciric ridge between the Ciric and Sapte Oameni streams. The regulation of the Bahlui is also a fundamental element. It has been related to the improvement of navigation on the Prut and the construction of a navigable canal from Ungheni (at the Bahlui-Prut confluence to Iasi). The scheme has also been associated with a diversion of water from the river Siret across the watershed at Strunga with hydro-electricity generation as the water descends eastwards to the Prut. However this plan of 1954, aiming at the expansion of heavy industry in Moldavia, has not yet been fulfilled. But by concentrating on high-density residential development on the terraces Iasi has experienced a rapid growth of population with relatively little increase in the built-up area.

CONSTRAINTS ON PLANNING

Despite central planning, controlling the growth of big cities remains difficult. (11) There has been a move towards comprehensive spatial planning for the 1980s and 1990s but the rigidity of such plans at the present time of technological change and economic recession is now seen as a serious drawback. All urban plans need to be flexible to cater for unforeseen developments. Enclaves of farmland (or waste land) may be left in cases when planned growth does not take place while growth beyond expectations can result in the occupation of green spaces and the generation of serious pollution problems. The expansion of Nowa Huta steelworks, planned in 1949 for 1 million tonnes annual production but subsequently enlarged to reach a capacity of 12 million tonnes, has meant that additional housing has encroached on a green belt separating the new industrial settlement from the rest of Krakow, which has itself expanded on to land originally earmarked for open space. Development envisaged under one medium-term plan may well be increased under a subsequent plan, and furthermore the power of the ministries in identifying sites for new projects is usually sufficient to overcome any constraints arising from a town plan. Local authorities are gradually obtaining greater powers to co-ordinate economic and physical planning but the problem of coping with centralized investors remains a delicate one throughout Eastern Europe.

Private ownership of land is significant. In the various settlements of Poland privately owned land accounts for 40.3 per cent of the total: in many individual settlements the proportion exceeds a half and in some of the largest cities it approaches 50 per cent (Krakow 48.7; Lodz 49.1). (12) If land is to be transferred to state ownership the rights of the private owner must be respected and compensation is payable; the situation varies from country to country but even in Romania where private citizens may hold the right to occupy pieces of land the need to compensate families for relinquishment of this right and the abandonment of their home (albeit at rates which are insufficient to allow for the building of a new privately owned house in lieu) is a cost and time factor in the redevelopment of the inner city. 'Though modest by standards in the West legal costs, compensation and higher prices for farmland can seriously erode the always inadequate funds for development which are at the disposal of city councils, ministries and industrial enterprises.' (13) So state organizations prefer to develop land which is already in public ownership. Thus comprehensive redevelopment can be seen in the old city area of Warsaw, which was nationalized in 1945. On the other hand privately owned land affords scope for development of individual houses separate from the apartment blocks erected by the state and other public bodies. Despite the existence of procedural controls such enterprise may be tolerated since it helps to satisfy the demand for more housing. Even in Romania where it is virtually impossible to erect a detached house in a town occasional irregularities may be observed.

Again there is official concern because excessive growth in certain cities may complicate food supply and inhibit the expansion of towns important for the future prosperity of other functional regions. Shopping facilities in some cities are highly unsatisfactory. Food shops in Prague amounted to only 122 square metres of space per thousand of the population (not taking account of all the commuters and visitors who use the city's shops) which compares unfavourably with other large cities in Czechoslovakia (Plzen 142 and Ostrava 146) and even smaller towns. Even normal provision is well below western standards: shopping space in department stores in Prague amounted to only 25.5 square metres per thousand of the population compared with more than 140 in Frankfurt am Main (FRG). Moreover in the early 1970s it was reported that housing shortages and poor transport discouraged migrants from moving to the city to ease the labour shortage. It was then stated that 200,000 people of working age would need to move into Prague by the end of the century. Equally the growth of Bratislava has been very rapid but not altogether harmonious with continuing housing shortages and inadequate infrastructure: transport has been unable to cope with the heavy demand while the water supply situation moved into a crisis in 1972 with the pollution of the Podunajeske Biskupice source by the Slovnaft plant.

However, despite the constraints, the planning of major cities is becoming more sophisticated and the following examples, dealing with Bratislava and Sofia, indicate the importance of site selection, functional specialization, economies of scale along with improved amenities and transport.

The development of Bratislava is planned to continue through to the year 2000 with apartment blocks constructed by building enterprises from various parts of Czechoslovakia. The central area of 700 ha will extend to both sides of the Danube, comprising several zones each with a distinct function and all linked by footpaths as well as roads. Transport facilities will congregate around the main railway station while the cultural zone, focusing on the Slovak National Theatre, will be close to the river. Suburban housing complexes are emerging with the largest, Paetrszalka, accommodating 150,000 people and justifying a fast railway link to the centre and a new bridge across the Danube, named after the Slovak National Uprising. Sofia cannot be confined and the planners are now providing for expansion to the 'natural limits': the Vitosha and Lyulin mountains to the south and the railway junction and Balkan Mountains to the north. Development will be most pronounced in the north-western and south-eastern directions and, conventionally for East European capital cities, will involve residential neighbourhoods, new parks, an enlarged forest belt (extending along the Iskar valley and including woodlands on the Lozenska, Lyulin, Plana, and Vitosha mountains) and a metro system ultimately consisting of 100 km of underground and surface lines. (14) Improvements in city infrastructure will have to be very substantial before western standards are reached. More reliable and better quality water supplies call for major new storages like the Bankya system for Sofia which will supplement the older supply systems in the Vitosha and Rila mountains.

THE CONCEPT OF CITY REGIONS

The logic of planning on the basis of city regions is now widely accepted. It is expressed in simple terms in the survey of macro-regions carried out by the UN Economic Commission for Europe in 1971 (Figure 6.4). These regions were built up from smaller administrative units (apart from Albania which was not subdivided and Yugoslavia where the republics were taken without modification), each with its regional city and a distinctive industrial profile. The regions also vary in the level of socio-economic advancement, measured here in terms of per capita GNP. Regions in Czechoslovakia and the GDR frequently approach (and exceed) the European average whereas several of the Balkan regions are relatively backward. In practice however planning exercises are much more sophisticated. The recognition of urban agglomerations connected by lines of communications which will comprise the growth axes of the future has been much discussed by Polish researchers led by K. Dziewonski. (15)

The approach is well illustrated by A. Potrykowska who lists a total of forty-seven functional urban regions in Poland, each consisting of a centre and a number of surrounding zones where commuting declines in intensity from high levels of more than half the active non-agricultural population to less than 10 per cent (Figure 6.5). (16) These regions vary in area and population: in two cases the population of the core is below 50,000 (Chelm and Zamosc) while in two others it exceeds 1 million (Katowice and Warsaw). And the strongest regions tend to attract commuters from weaker adjacent regions: supraregional centres can be identified in Warsaw, drawing commuters from Lomza-Ostroleka, Plock, Siedlce, and Radom; and

211

Figure 6.4 Macro-regions in Eastern Europe

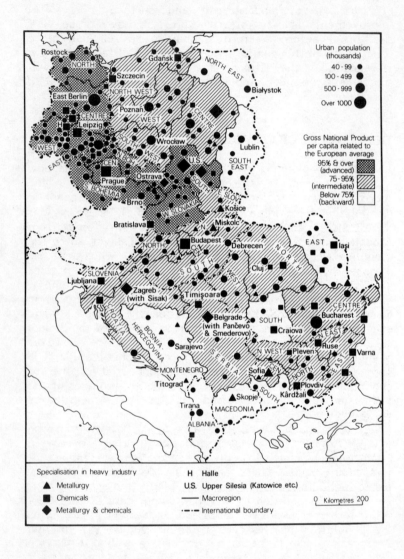

Sources: Based on the UN Economic Commission for Europe's 'Economic survey of Europe 1971' and statistical yearbooks.

Figure 6.5 Regional planning in Poland

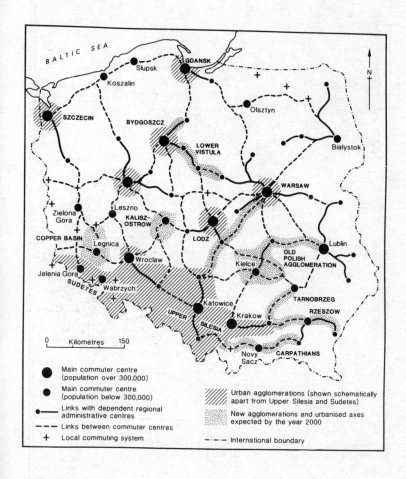

Sources: Based on Polish researches listed in notes 15 and 16 (p.224-5) especially A. Potrykowska (1985).

213

Figure 6.6 Administrative regions of Eastern Europe

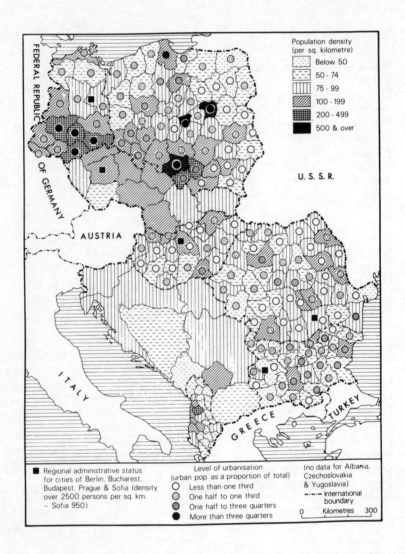

Source: Statistical yearbooks.

large provincial cities: Bialystok, drawing commuters from Lomza-Ostroleka and Suwalki: Gdansk from Elblag; Katowice from Bielsko-Biala and Czestochowa; Kielce from Ostrowiec Swietokrzyski; Krakow from Tarnow; Lodz from Piotrkow Trybunalski; Lublin from Chelm, Siedlce and Zamosc; Poznan from Gniezno, Konin, and Pila; Rzeszow from Krosno and Przemysl; Szczecin from Gorzow Wielkopolski and Wroclaw from Opole. In almost all cases the population of the core exceeds 200,000 and the balance of employment and residence suggests a well-developed labour market. With a hierarchy clearly established planners can then plan new investment in industry and infrastructure to strengthen the weaker centres so that they realize their potential. As the number of administrative regions increased to forty-nine, 'cities which then became new seats of regional government immediately started to press for the same service functions which earlier centres had already obtained'. (17) The cities may be built into growth axes which comprise a series of corridors intersecting at Warsaw: Szczecin to Lublin, Gdansk to Przemysl, Olsztyn to Krakow, Suwalki to Katowice, Bialystok to Walbrzych and Siedlce to Poznan along with the circuit Szczecin-Gdansk-Bialystok-Lublin-Krakow-Katowice-Wroclaw-Zielona Gora-Szczecin and the axis from Gdansk to Plock, Lodz, Katowice and Bielsko Biala. (18)

Settlement planning

It is now usual in Eastern Europe to use a regional administrative system in which the central towns are some 75 km apart (Figure 6.6). Where industrialization is not already advanced there is sufficient infrastructure and political support to introduce new industries and mobilize the surplus population. Planners in the GDR give the highest priority to the capital and the regional cities. A quarter of the population of the country is within twenty minutes travelling time of the centres of these cities and a further fifth is within one hour's travel. (19) Migration into the cities and their commuting zones is encouraged by the provision of jobs and housing. Berentsen has noted a +0.92 correlation between net migration and change in housing availability in the Dresden area between 1963 and 1967. (20) This is done at the expense of regional equality in the quantity and quality of housing since high-quality housing is highly concentrated in such growth centres as Eisenhuttenstadt, Halle-Neustadt, Rostock, Schwedt, and Stralsund. There has been no sustained industrialization drive in the north but various light industries (often related to agriculture) have been established in the main regional centres (Neubrandenburg, Rostock, and Schwerin) and the range of employment has increased especially for rural dwellers who are prepared to commute. The government's policies have been very successful although people were slow to move into Halle-Neustadt at first because of limited services and a high crime rate. The commuting hinterlands of the fifteen regional (Bezirk) cities overlap with those of the smaller towns. But the 140 district (Kreis) towns also feature prominently in planning as 'the decisive centres of everyday life'. Only a fifth of the population is more than forty minutes' travel from a district or regional town (and only 4 per cent are more than an hour away).

In Czechoslovakia the planners are concentrating on cities with potential to develop to population levels of 50,000. Furthermore towns may be grouped into agglomerations with a minimum population of 100,000: nineteen have been identified. With specialization between the various towns the group will function as an effective urban core capable of extending influence over an extensive rural hinterland. Thus it will not be necessary for small towns to proliferate in order to overcome the antagonism between the dynamic city centre and the remote countryside. In 1971 the Hungarian government adopted the National Settlement Development Strategy which has become the master plan for the country's urban development. (21) This may be seen as the spatial component of the New Economic Mechanism, providing a stimulus for regional economic development. Whereas previously investment had been concentrated on Budapest and a number of new towns concerned with extractive industries and the manufacture of chemical and metallurgical products at convenient raw material collecting centres (Dunaujvaros, Kazincbarcika, Komlo, Oroszlany, and Varpalota) there was now a conception of a settlement system involving eight levels below the state capital.

Particular importance has been attached to the cities which make up the three highest orders: a group of five large regional cities (Debrecen, Gyor, Miskolc, Pecs, and Szeged) which offer the best potential as counterweights to

Budapest; a group of six slightly less prominent regional centres (Bekescsaba, Kecskemet, Nyiregyhaza, Szekesfehervar, Szolnok, and Szombathely), and a further group of twelve centres covering the remaining county towns along with four other centres of similar population size and economic importance. By the early 1980s considerable progress had been achieved with all the main regional centres but one (Gyor) having reached the minimum population target of 150,000 and all cities in the next group having passed their designated minimum of 80,000 apart from Bekescsaba and Szolnok. These targets have been reached partly because of a higher level of natural increase and partly through the diversion of migration currents leading to Budapest. In Hungary preferential growth axes have been identified on a north-south alignment between Gyor, Budapest, and Pecs and between Miskolc, Szolnok, and Szeged and on a complementary east-west alignment from Nyiregyhaza to Debrecen, Szolnok, Szekesfehervar, and Szombathely. It is not possible to identify studies for all East European countries but it would appear that a similar logic is being employed in the Balkans. (22)

Decentralization: migration controls and the relocation of industry

A particularly relevant issue in urban planning is the containment of large cities in order to minimize congestion and simultaneously enhance the growth prospects of smaller towns and backward regions. Rationalization of industry, with greater integration and a larger scale of production, is conducive to decentralization, although the attraction of new resources, like the brown coal of Konin, and the contribution of road transport in enhancing the potential of towns lacking good railway services, is significant under any political system. (23) But even in the planned economy the forces of agglomeration remain strong:

> while increases in output may be achieved more cheaply and easily by expanding existing factories than by erecting new ones in areas which have no tradition of such production, and while the true costs of concentration are obscured by the subsidy of fares on public transport, of rents and by the absence of a market to determine the true value of land in cities,

217

the national will to spread industry and urban life more widely may be frustrated by local piecemeal action of the various industrial ministries and their constituent enterprises. (24)

The desirability of relocation may be questioned if industrial efficiency, mainly in terms of productivity, is appreciably higher. It may therefore be desirable to decentralize labour-intensive industries while expanding the high-technology industries which can best thrive in an urban environment with an industrial tradition and a good infrastructure. This strategy would seem to strike a balance between the economic imperative of efficiency so basic to the achievement of high living standards and the social argument against excessive concentration.

In Warsaw there has been some relocation of industry but it has affected the surrounding areas most of all and has not prevented the further industrial growth of the city itself; partly because of economic efficiency and partly for political reasons, to increase the prominence of the industrial proletariat. However the amount of expansion in Warsaw was carefully controlled: immigration was severely restricted after 1954 and additional jobs were provided in the small towns of the region through the transfer of certain firms from the capital. The scale of movement has been less than was anticipated and some new investments (including a new steel-works) initially allocated to small towns have been put into Warsaw itself, but nevertheless a balance has been struck. This has been beneficial to the smaller towns and also to Warsaw for the reduced pressure from industry has simplified the redevelopment of mixed housing-industrial areas like Powile and Sielce damaged in the war. Much industry remains in Praga and Wola (key centres of industrial employment for the east and west of the city respectively) and new industrial estates have been laid out at Mlociny, Sluzewiec, and Zeran. Meanwhile the decongestion exercise in Upper Silesia has created new jobs in the areas immediately adjacent to the old industrial zone. (25)

Decentralization may also be approached by controls on migration but this policy has not been a success in Poland: problems of labour shortage and underused factory capacity have led to its abandonment. But the strategy is still used in other countries. The Bulgarians have controlled migration into Burgas, Plovdiv, Ruse, Sofia, and Varna since 1955

(Pleven since 1956) through the issue of residence permits. And restrictions were extended to all but 32 of the country's 174 towns in 1974: the exceptions relate to small towns with a declining population or new towns where large industries have recently been started. But the policy has been implemented only with difficulty. It is common for marriages to be arranged between residents and non-residents in order to facilitate movement into the big cities and evasion of the law has been widespread, with 120,000 illegal residents reported in Sofia alone. It seems that the inexorable growth of the economy in the large cities requires substantial migration and the grave shortage of construction workers has led to the issue of permits on numerous occasions. Such workers were welcome in Sofia in 1961 when the Kremikovtsi metallurgical combine was under construction and further recruitment has taken place in 1967 and 1973 with migrants being required to give pledges (for five and ten years respectively) that they would work in the construction industry. While it is appropriate to ensure that the population does not race ahead of job opportunities economies of scale cannot be sacrificed by a stagnant population.

The migration policy has been complemented by an effort to expand industry in small towns and also in the villages. Some authors therefore consider that the diversion of industrial growth from the larger to the smaller urban settlements has been more successful under socialism than it would probably have been under capitalism. But decentralization from capital cities is never easy because influential individuals and powerful organizations can use their 'clout' and ingenuity to counter administrative measures. But there are genuine economic problems as well. Experience in Sofia is highly revealing. Attempts to move enterprises out of Sofia met with only limited success during the early 1980s despite the grave housing and labour shortages. In the course of 1981 only ten of the twenty-three organizations listed for relocation actually moved out of the Bulgarian capital. Institutes like those dealing with soil and viticulture need not be situated in the capital city but it seems that some have altered their functions in order not to appear anomalous. Thus an enterprise servicing agricultural machinery due to move out of Sofia to the small town of Isperih has avoided the transfer by transforming itself into an enterprise for servicing and marketing. And although the Textilna Slava textile factory was closed down

219

according to the relocation plans the saving on labour in Sofia was not achieved because its production was allocated to other textile enterprises in the city. The plant for secondary raw materials of non-ferrous metallurgy became a department of the Gara Eliseyna metallurgical plant and was formally recognized as having moved out of Sofia, although work continued to be done in the old premises. Some relocation orders have been withdrawn like that for the Petko Napetov paper factory, where modernization of old equipment had started. Close connections between personnel in enterprises and ministries can obviously be important when special dispensations are forthcoming.

The case of Hungary

In Hungary the need for decentralization has been all the greater in view of the dominant position of city in national life. The rapid growth of Budapest in the late nineteenth century as an imperial city followed by the setting of relatively modest frontiers under the Treaty of Trianon has prevented the emergence of any strong counter-influence. While Budapest's population has risen substantially both in absolute terms and as a share of the total population of Hungary the share of the total urban population has fallen slightly from 43.3 per cent in 1930 to 41.6 in 1960 and then more sharply to 38.9 in 1970 and 36.2 in 1980. Quite a marked change occurred during the 1960s with reduced rural-urban migration to Budapest and increased movement into other towns. Thus between 1960 and 1963 there was a net migration away from the countryside of 191.9 thousands: 88.1 thousands (45.9 per cent) to Budapest and 103.8 thousands to other towns. During the next four years the net migration loss to the countryside was 157.3 thousands: 47.9 thousands (30.5 per cent) to Budapest and 109.4 to other towns. Then between 1968 and 1971 the rural areas lost 138.8 thousands: 34.5 to Budapest (24.9 per cent) and 104.3 to other towns.

Relocation may create difficulties for enterprises which need to retain close links with the capital. Although distances may not be great the transport system is not always efficient enough:

under such conditions intimate contacts with laboratories and research personnel and close technical co-operation between enterprises, crucial in growth industries, are very difficult to maintain without a metropolitan location; for the most valuable industries of Budapest, and, to a large extent of Hungary as a whole, metropolitan location today seems essential. (26)

Access to cheaper (though less well-qualified) labour in the provinces does not seem to be regarded as adequate compensation. Industries which do move out tend to be those which generate serious pollution problems or require a great deal of space. 'In general the firms equip their relocated units mostly with worn-out amortised machinery from their metropolitan establishments which then receive more advanced or less polluting or otherwise obnoxious equipment'. (27) Perhaps higher education could be strengthened in the provinces at the expense of Budapest where almost half of all students in higher education are studying. Also research activities concerned with agriculture, food processing, and mining could quite easily be relocated.

Alternative locations for industry are identified by the national settlement plan which is very highly formalized in the Hungarian case. However the official hierarchy may not be entirely appropriate for there is always a risk of political manipulation to produce more centres than economic and social considerations fully justify. Compton argues that while the upper tier of five centres is entirely justifiable (although infrastructure needs much attention: only Szeged and Szolnok have adequate water supplies) the intensive development of the eighteen places comprising the next two levels in the hierarchy involves unnecessary duplication. (28) At the same time it is suggested that the emphasis given to the three categories of regional centres, while reducing the further growth of Budapest, has a negative effect on the small towns and the villages, to say nothing of the settlements with fewer than 3,000 inhabitants left outside the strategy altogether on the grounds that they lack potential for development. Finally on a different note it is evident that changed economic circumstances may prevent the allocation of resources at the planned levels and introduce serious inefficiencies through the need for a lengthy gestation period.

So although Budapest's share of employment in industry has fallen to just under one-third and will continue to fall in future the city's superior productivity record will ensure that there is further expansion of output in absolute terms, with particular emphasis on high-technology industries. A new geography of industry is thus emerging in Budapest and this can be seen not only in structural terms but also spatially. Compton points out that in the past the policy was to concentrate industry in the south-eastern part of the city to the leeward side of the prevailing winds. But now new residential neighbourhoods have moved into these areas and the present policy aims at eliminating industrial-residential neighbourhoods by creating western-style industrial estates in all parts of the city. While several districts will see the retention and extension of their industrial function: (Angyafold, Csepel, Ferencvaros, Kelenfold, Kobanya, Obuda, and Ujpest) new estates will emerge in Nagyteteny, Pestujhely, Rakospalota, Rakosvolgy, and Soroksar. Some industrial development will be allowed in open country immediately beyond the city boundary on land originally designated as green belt (in Budaors and Szazhalombatta for example). Meanwhile a good deal of industry has been displaced from such inner suburbs as Obuda and Jozsefvaros and it is likely that eventually only some small labour-intensive workshops will be left in the more central areas of the city. This industrial strategy goes hand in hand with slum clearance projects in the inner city and the building of new residential neighbourhoods on agricultural land close to the city boundary. This has obvious implications for transport and services. A rapid-transit rail system, mostly underground, will be complemented by upgraded suburban surface railways and some express tramways; while the old monocentric structure with administration, commerce, and cultural facilities concentrated in the city centre will be modified by the construction of new service centres, and especially six major local centres forming a ring round the present commercial core: Buda (Moszkva ter), Kispest, Lagymanyos, Obuda, Ujpest, and Zuglo.

CONCLUSION

Planning the settlement pattern is thus a highly complex operation. Growth should not take place on an excessively polarized basis and hence the concern over optimum population size for the leading towns. However the consensus has been a shifting one with the maximum population moving upwards from 100,000 to some 300,000 inhabitants. (29) But planners now find it very difficult to generalize and, since the 1970s, preoccupation with optimum size has been restrained. Instead development costs need to be examined in each individual case and judgements over the priorities to be accorded to different functional regions and the various towns within each regional system have to be made pragmatically. The other side of the coin concerns the reception areas for the overspill. It is no use trying to distribute population and industry too widely for 'the peppering of backward areas with large numbers of small projects has come to be regarded as a recipe for failure'. Hoffman suggests that for Yugoslavia there is an ideal balance that ought to be struck, so as to create a network of strong dynamic industrial centres which can effectively modernize the entire countryside. (30) However Slovenia fails to emulate this model because of the highly decentralized character of urban development: nearly two-thirds of the population live in settlements with a population below 2,000 and 'there does not exist a single large enterprise which might serve as a development pole of sufficient strength to leave its mark on the structure of the republic as a whole'. (31) By contrast 70 per cent of post-war investment in Montenegro has been concentrated in three large enterprises located in two cities: spread effects have been relatively weak. Each large city and each city region requires an individual assessment.

NOTES

1.   M. Dobrowolska and J. Herma (1968) 'Migration of manpower in South Poland', Geographia Polonica 14: 321-31. See also I. Abrams and R. Francaviglia (1975) 'Urban planning in Poland today', Journal American Institute of Planners 41: 258-69; A.H. Dawson (1984) 'Poland: thirty year rural-urban transition', Cities 1: 175-84; M. Dobrowolska (1976) 'The growth pole concept and the socioeconomic development of a region undergoing industrialization', Geographia Polonica 33(2): 83-101; P. Eberhardt (1980) 'Settlement concentration and industrial productivity in Poland', Geographia Polonica 43: 231-50; A.J. Francis (1971) 'Urbanisation process in Poland', Geography 56: 133-5.
2.   J. Friedrichs (1988) 'Large cities in Eastern Europe', in M. Dogan and J.D. Kasarda (eds) The Metropolis Era, London: Sage, 128-54; L. Strasiewicz (1969) 'Capitals of

the socialist countries in Europe', Geographia Polonica 16: 27-40.

3.   L. Dienes (1973) 'Urban growth and spatial planning in Hungary', Tijdschrift voor Economische en Sociale Geografie 64: 24-38 (quote p. 28).

4.   F.W. Carter (1974) 'Concentrated Prague', Geographical Magazine 46: 537-44; P. Mateju et al. (1979) 'Social structure spatial structure and problems of urban research: the example of Prague', International Journal of Urban and Regional Research 3: 181-202; J. Musil (1968) 'The development of Prague's ecological structure', in R.E. Pahl (ed.) Readings in Urban Sociology, Oxford: Pergamon, 232-59.

5.   F.W. Carter (1986) 'City profile: Tirana', Cities 3: 270-81.

6.   C. Herbst (1968) 'Bucharest: main industrial city of Romania', Revue roumaine: geographie 12: 133-6.

7.   French and Hamilton (1979) 381.

8.   R.D. Asmus (1983) 'Jena: an atypical city in the GDR?', Radio Free Europe Background Report 76.

9.   E. Lettrich (1971) 'Kecskemet: a typical town of the Great Hungarian Plain', in Sarfalvi (ed.) 145-61; H. Szulc (1972) 'Village relict features within the spatial layout of some Polish towns', Geographia Polonica 24: 241-54.

10.   D. Turnock (1989) 'Housing in Romania', in Sillince (ed.).

11.   R.J. Fuchs (1980) 'Urban change in Eastern Europe: the limits to planning', Urban Geography 1: 81-94.

12.   French and Hamilton (1979) 208.

13.   ibid. 210.

14.   F.W. Carter (1973) 'Post-war functional and structural changes within the Sofia conurbation', University College London Department of Geography Occasional Paper 21.

15.   K. Dziewonski (1973) 'The geographical differentiation of contemporary urbanization', Geographia Polonica 27: 31-43; K. Dziewonski (1975) 'The place of the urban agglomeration in the settlement system of Poland', Geographia Polonica 30: 9-19; K. Dziewonski (1976) 'Changes in the process of industrialization and urbanization', Geographia Polonica 33(2); 39-58; K. Dziewonski (1978) 'Structure population and future settlement system in Poland', in Hansen (ed.) 145-54; K. Dziewonski et al. (1984) 'The Polish settlement system', in Bourne et al. (eds) 360-76; K. Dziewonski and M. Jercynski

(1975) 'Urbanization in Poland', in R. Jones (ed.) Essays on World Urbanization, London: Philip, 263-77.

16. M. Jerczynski (1977) 'Urban system of Poland: its development and functional organization', Geographia Polonica 37: 73-88; M. Jerczynski (1981) 'Development of the national system of cities as related to migration', Geographia Polonica 44: 97-109; A. Potrykowska (1985) 'Spatial structure of functional urban regions in Poland', Geographia Polonica 51: 113-26; G. Weclawowicz (1981) 'Towards a theory of intra-urban structures of Polish cities', Geographia Polonica 44: 179-200; A. Zagozdzon (1976) 'Regional and subregional centres in Poland: a general characterization', Geographia Polonica 32: 63-74.

17. French and Hamilton (1979) 369.

18. A.H. Dawson (1987) 'Transport and the pattern of settlement in Poland: the impact of post-war policies', in Tismer et al. (eds) 306-27; G. Kind (1981) 'Modelling of settlement systems for regional planning', Geographia Polonica 44: 33-43; T. Lijewski (1980) 'The centrality of towns as reflected by transport indices', Geographia Polonica 43: 251-8; B. Malisz (1975) 'The band-node model of the settlement network of Poland', Geographia Polonica 32: 121-32.

19. W.H. Berentsen (1980) 'Spatial pattern of retail sales per capita in the GDR and East Berlin', Die Erde 111: 293-300; F. Grimm (1981) 'An investigation of the settlement system of the GDR', Geographia Polonica 44: 151-62; H. Ludeman and J. Heinzmann (1978) 'On the settlement system of the GDR', in Hansen (ed.) 121-43; G. Mohs (1975) 'Society and settlement: developing trends in spatial structure in the GDR', Geographia Polonica 30: 21-7; K. Scherf (1981) 'Basic aspects of settlement structure policy in the GDR', Geographia Polonica 44: 60-3.

20. W.H. Berentsen (1981) 'Regional change in the GDR', Annals Association of American Geographers 71: 50-66.

21. I.T. Berend (1974) 'Development strategy and urbanisation in Hungary', in A. Brown et al. (eds) 271-86; L. Dienes (1973) 'Urban growth and spatial planning in Hungary', Tijdschrift voor Economische en Sociale Geografie 64: 24-38; J.A.A. Sillince (1987) 'Regional policy in Hungary: objectives and achievements', Transactions Institute of British Geographers 12: 451-64.

22. P. Deica and I. Stefanescu (1972) 'Form of the territorial grouping of the settlement network in the SRR',

Revue roumaine: geographie 16: 215-16; P. Deica et al. (1976) 'The interdependence between the territorial structure of the national economy and the unitary system of settlements in Romania', Revue roumaine: geographie 20: 203-11; I. Iordan (1975) 'Commutation in the periurban zone of Bucharest', Revue roumaine: geographie 19: 87-94; Musoroiu (1974); L. Panaite et al. (1978) 'First rank polarizing centres in Romania's urban settlement network', Revue roumaine: geographie 22: 251-6; Ronnas (1984).

23. A. Dawson (1979) 'Factories and cities in Poland', in French and Hamilton (eds) 349-86. See also J. Debski (1974) 'The spatial structure of the Gdansk agglomeration', Geographia Polonica 28: 127-44; A. Kuklinski (1967) 'Changes in the regional structure of industry in People's Poland', Geographia Polonica 11: 97-109; S. Leszczycki (1975) 'The growth limits of urban-industrial agglomerations in spatial development at the national level', Geographia Polonica 32: 105-11.

24. A. Dawson (1979) 'Factories and cities in Poland', in French and Hamilton (eds) 366.

25. B. Kortus (1972) 'The effect of industry on the development of towns: selective problems in the case of Cracow', Geographia Polonica 27: 183-9; B. Kortus (1975) 'Structure and development trends in the Cracow agglomeration', Geographia Polonica 30: 113-24; L. Pakula (1980) 'The Upper Silesian core region: growth and evolution', Geographia Polonica 43: 213-22.

26. L. Dienes (1973) 'Urban growth and spatial planning in Hungary', Tijdschrift voor Economische en Sociale Geografie 64: 32. See also Bencze and Tajti (1972); L. Dienes (1973) 'The Budapest agglomeration and Hungarian industry: a spatial dilemma', Geographical Review 63: 356-77.

27. L. Dienes (1973) 'Urban growth and spatial planning in Hungary', Tijdschrift voor Economische en Sociale Geografie 64: 32.

28. P.A. Compton (1984) 'Hungary: the national settlement development strategy', Cities 1: 374-84.

29. French and Hamilton (1979) 11.

30. G.W. Hoffman (1973) 'Currents in Yugoslavia: migration and social change', Problems of Communism 22(6): 16-31.

31. ibid. 24.

Chapter Seven

CONTINUITY AND CHANGE IN THE RURAL AREAS

Although rural-urban migration is occurring everywhere in Eastern Europe it is in the Balkans where the greatest transformation is now taking place. Some depopulation was evident in Slovenia in the late nineteenth century but in many parts of the Balkans absolute decreases of population did not occur until the post-war era. But formerly self-sufficient peasant communities have now been radically affected by modernization and the towns have begun to assimilate rural dwellers in unprecedented numbers. In turn the villages have been forced to adjust to the break-up of kinship groups and the decline in population. (1) Agriculture has been much affected by the reduction in population pressure: it is much more highly commercialized and offers regular employment rather than a way of life. However the efficiency of co-operative and state farms has frequently been a matter of great concern and Eastern Europe as a whole is no longer a net exporter of agricultural products. At the same time the peasant character of farming is retained, especially in Poland and Yugoslavia where collectivization was never pursued to extremes by the state. In other countries some individual farms remain, usually in areas where co-operation would be extremely difficult to organize, but constraints of various kinds (prescribed areal limits, rising taxation levels, or prohibition of employment of non-family labour) prevent the emergence of large capitalist farms. There is also some scope for private enterprise on the small family plots allocated to members of co-operative farms.

227

Figure 7.1 Urbanization and inter-regional migration in Romania

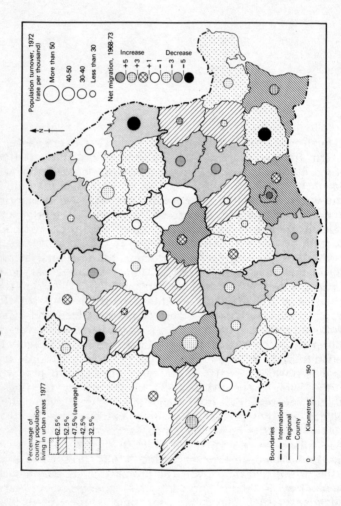

Percentage of
county population
living in urban areas 1977

62.5%
52.5%
47.5% (average)
42.5%
32.5%

Population turnover, 1972
(rate per thousand)

More than 50
40-50
30-40
Less than 30

Net migration, 1968-73

Increase   Decrease
+5 +3 +1 -1 -3 -5

Boundaries
International
Regional
County

0   150
Kilometres

Source: Statistical yearbooks.

THE RURAL PROBLEM

Given the prevailing western phenomenon of counter-urbanization underpinned by preference for the rural milieu by many commuters and a substantial movement into the countryside by retired people it may seem inappropriate to project rural parts of Eastern Europe as backward areas in contrast to the dynamism of the towns. Certainly the country districts are much less isolated nowadays although the substantial volume of daily commuting to work is generated by somewhat different considerations from those which apply in the west. There is some recognition of urban regions or metropolitan areas where commuters have forged an intimate link between a city and its immediate hinterland.

Even where commuting is not strong the rural areas may not be altogether deprived of non-agricultural income. Mining, quarrying, textile manufacture, and food processing frequently occur in rural areas. Visits to the towns for commercial and social reasons are commonplace. Mihailovic observes that peasant girls in Czechoslovakia like to marry in the towns and that this preference has become a veritable obsession in Yugoslavia. (2) However it is increasingly evident that the more affluent families, with resources and influence to exercise a modicum of choice as to their place of residence, tend to prefer the convenience and relative sophistication of urban life. At the same time the young people frequently leave the family home in the countryside, perceiving greater opportunities in the towns. Figure 7.1 deals with urbanization and migration in the regions of Romania and it is evident that out-migration (across regional boundaries) is often very prominent in regions with a low level of urbanization whereas in regions which are highly urbanized there is greater scope for short-wave rural-urban migration within the administrative limits. This migration process is socially disruptive and can bring about the ruralization of towns which face a high level of in-migration. What is wanted is an extension of the influence of the towns (economically and culturally) to achieve an urbanization of the villages so that differences in per capita income are reduced and country people can use the amenities of the towns more freely. (3)

The situation has become particularly unsatisfactory because in the early post-war years rural planning was very limited as priority was given to the development of

229

industrial complexes in the towns. The formation of state and collective/co-operative farms resulted in some rural investment with agricultural buildings (for livestock, storage, and vehicle maintenance) provided in many villages, especially the larger ones, on an ad-hoc basis. But rural industry tended to decline. Some losses of rural industry have been purely technical, arising out of administrative changes. At the same time Tatai acknowledges that in the 1950s artisans providing rural services were squeezed out of the economy by a discriminative fiscal policy. (4) Food processing became much more concentrated in the 1950s and 1960s and the small village distilleries, mills, dairies, and bakeries closed down. This gave economies of scale in production but high distribution costs at the same time. There were also closures of brick works, coal-mines, and timber yards. Much of the rural employment growth was restricted to the larger villages so that 'teachers in the upper grades of the general school, greengrocers, butchers, shoe repairmen and barbers left the smaller places of their own volition and moved into the higher-order centres'. (5) Significant differences in urban and rural living standards arose because many rural households had no regular income. But more important perhaps was the difference in culture between the sophistication of the towns and the traditionalism of the villages with agriculture as a way of life (through private gardens, plots, or small farms) allied with continuing religious observance and popular culture.

EXTENDING THE URBAN NETWORK

Even the smaller towns experienced some neglect. Wartime damage to buildings and the killing of Jews, who often comprised an important part of the business community, created serious set-backs even before nationalization and central planning. However problems have increased with further industrialization because larger enterprises require the infrastructure which only a large town can offer. Normally any loss of employment in handicrafts can be compensated for by growth in the service sector, geared to a rural population which shares in the economic development process working towards higher incomes. In Eastern Europe, however, squeezing of the rural sector to generate wealth for investment in large industries has postponed the expansion of the tertiary sector in small towns. Many of them have stagnated and have been highly dependent on the dispersal of manufacturing which, according to Mihailovic, 'became a sort of ideology in Bulgaria'. (6) He goes on to say that eventually the problems will be solved through a higher level of industrialization and the diversion of more resources to agriculture. But over-concentration in the big cities will have to be avoided and problems may still remain for small towns situated away from the main growth corridors. (7)

The problems of micro-urbanization are very evident in the northern and western territories of Poland. There was a lack of demand for services as the formation of the first collective farms led to the establishment of new machinery and maintenance facilities outside the traditional centres. Deliveries to state warehouses meant that the peasants had little produce to take to local markets. Meanwhile there was a reduction in personnel providing services since many of the craftsmen, persecuted by high taxation and administrative pressures, tried to escape to the large industrial enterprises which seemed to offer a secure living. On the top of all this the fact that many of the peasant farmers were resettled from the east where they had become used to relatively primitive farming methods and a prime objective in self-sufficiency made the decline in local marketing all the more striking. In the Bydgoszcz and Poznan areas strong traditional ties between town and village were significant but in all parts of the country, to a greater or lesser extent, the removal of pressures to collectivize in 1956, meant the building of local commercial structure afresh. (8)

The situation has been eased by the formulation of policies to restrain the growth of the capital city and pump more investment into growth centres further down the central place hierarchy. Accelerated growth of the main provincial cities is an obvious policy if investment and migration flows are to be diverted away from the capital and such a policy has been broadly successful. But small towns within the sphere of influence of the major cities have also benefited from decentralization and the problem of such towns in the Katowice, Krakow, Szczecin, Warsaw, and Wroclaw areas of Poland has been solved. Moreover information on comparative development costs suggests that small towns in more backward areas could expand. The cost of urban expansion can vary considerably and the initial population size is only one of several variables. Because of

Table 7.1 Settlements and migration in Poland 1950-80

Category		A	$\frac{1950}{B}$	C	A	$\frac{1960}{B}$	C
Poland	Total		24.82			29.56	
	Rural		15.74	63.4		15.38	52.0
	Urban	706	9.08	36.6	889	14.18	48.0
Small towns	Total	628	3.21	35.4	778	4.69	33.1
	Below 5,000	393	1.07	11.7	411	1.20	8.5
	5,000–20,000	235	2.15	23.7	367	3.48	24.6
Medium towns		50	1.52	16.8	69	2.10	14.8
Large towns	Total	28	4.34	47.8	42	7.39	52.1
	50,000–100,000	12	0.83	9.2	20	1.29	9.1
	0.10–1.00 million	16	3.51	38.6	21	4.96	35.0
	Above 1 million				1	1.14	8.0

A	1971		1980			D	Migration 1971-5		
	B	C	A	B	C		E	F	G
	32.81			35.73		+1,193.3			
	15.85	48.3		14.41	40.3	+461.7	−888.7		
889	16.95	51.7	804	21.33	59.7	+731.6	+888.7	0.82	1.00
741	4.91	29.0	612	4.41	20.7	+219.5	+158.6	1.38	1.24
359	1.10	6.5	263	0.78	3.7	+37.7	+9.2	4.10	2.24
382	3.81	22.5	349	3.62	17.0	+181.8	+149.4	1.22	1.10
97	2.92	17.2	114	3.52	16.5	+164.4	+205.0	0.80	0.75
51	9.12	53.8	78	13.40	62.8	+347.6	+525.1	0.66	1.00
27	1.87	11.0	40	2.71	12.7	+100.5	+146.5	0.69	0.72
23	5.93	35.0	37	9.07	42.5 ⎫	+247.1	+378.6	0.65	1.11
1	1.32	7.8	1	1.61	7.6 ⎬				

Sources: M. Jerczynski et al. (1981) 'Development of the national system of cities as related to migration', Geographia Polonica 44: 97-109; Statistical yearbooks.

Notes:
A Number of towns
B Population (millions)
C Percentage of urban population (and rural-urban ratio)
D Natural increase (thousands)
E Net migration (thousands)
F Growth by natural increase in relation to growth by migration
G Share of urban population 1970 related to the share of growth 1971-5

Table 7.2 Growth of the urban population and non-agricultural employment in Poland 1970-84

Regional (voivodship) groups	Urban population		Non-agricultural employment					
	1970-78	1978-84	1970-78			1978-84		
			a	b	c	a	b	c
With large urban agglomerations*	1.7	1.2	2.1	1.1	3.0	-1.0	-1.7	-0.4
With urban population over 40 per cent	2.3	2.2	2.8	2.3	3.2	-0.2	-1.6	0.9
With urban population between 30 and 40 per cent	2.4	2.5	3.5	3.6	3.4	0.0	-0.6	1.3
With urban population below 30 per cent	3.2	3.2	4.5	5.4	4.0	1.0	-0.2	2.0
Poland	2.1	1.7	2.7	2.1	3.2	-0.3	-1.4	0.5

Source: Statistical yearbooks.

Notes: a Total   b Industry   c Services
   * Gdansk, Katowice, Krakow, Lodz, Poznan, Szczecin, Warsaw, Wroclaw

their relatively low settlement densities and their relatively poor infrastructure cities of Ottoman or Imperial Russian origin tend to be more expensive to develop than cities situated in the former Prussian and Habsburg empires, irrespective of the total population size. Again development costs can vary considerably between sites lying on a preferential axis and land occupying the interstices between the main highways which could be more expensive to service: hence the preference in some cases for distinctly linear developments along a major transport axis. Lengthy journeys to work are a possible hazard but construction of a local industrial estate could solve this problem and simultaneously allow advantage to be taken of lower land values on the edge of the city in those countries which, like Poland, allow some discrimination in land values and rents. On the other hand the cost of development in a small town may be relatively low if economies of scale in building happen to coincide with spare capacity in the local services and good access to foodstuffs on farms in the immediate vicinity. Harmonizing the physical planning with economic planning can be difficult in small towns but where a large city enterprise is encountering difficulty in attracting suitable labour a branch plant in a small town with a good catchment may be an attractive proposition. (9)

It is a matter for some satisfaction that in Poland the growth of population and non-agricultural employment was by no means restricted to the main urban agglomerations during the 1960s. With 38.3 per cent of the population in 1960 the urban agglomerations took 65.3 per cent of the growth during the following decade, to increase from 11.38 to 13.22 million. Meanwhile the rest of the country registered a growth of 0.98 million (from 18.39 to 19.37). However the non-agricultural population showed a more rapid increase outside the agglomerations (35.2 per cent, from 7.80 to 10.55 million) than within (20.3 per cent, from 10.32 to 12.41 millions). (10) Rural-urban migration has been widely distributed through the hierarchy (Table 7.1) and the smaller towns have continued to grow (Table 7.2) as the government has tried to expand the tertiary sector to the level of advanced regions in order to check the emigration of the young. It is easy to draw people away by creating attractive opportunities elsewhere, but less easy to ensure adequate demographic resources in the exporting (backward) areas where age and sex structures can be seriously disturbed. Four-fifths of migration cases investigated in a

235

Polish study involved changes in work-place or occupation. Enterprises can help to stabilize population in areas around small towns. In Yugoslavia enterprises may assist commuters in various ways. People living a long way from the factory may be given subsidies to cover much of the travel cost. Alternatively the enterprise may build flats close to the factory and make them available to employees living a long way out of town (or financial help may be given towards the cost of building a private house nearer to the factory). Assistance towards the cost of secondary education may be given to young people who are expected to work for the enterprise eventually. Such stimulants have been particularly common in labour-deficit areas like Slovenia. (11) In some cases there is a particularly high level of integration at the level of the district; for example at Sonneberg on the southern frontier of the GDR where commuting for work in the toy factories (the speciality of this district) is entirely contained within the administrative limits of the Kreis. (12)

NEW TOWNS

The modernization of the rural areas has been achieved most dramatically by the building of new towns. The utopian aspects of socialism might well generate an imperative for new town development but the East European response to the 'garden city' movement in the west has been distinctly muted. The emphasis given to industrial growth has made it difficult to justify the heavy investment needed for the creation of new towns and the slow progress towards settlement planning has imposed a further constraint. Only under the special circumstances of acute congestion in existing cities and overwhelming industrial potential in specific rural areas have 'overnight' transformations occurred. Hamilton suggests that there are three types of new town situation. Some large new towns have been built to comprise integral parts of existing cities like Halle-Neustadt in the GDR, Havirov and Poruba near Ostrava in Czechoslovakia, Nova Huta near Krakow in Poland, and Novi Beograd in Yugoslavia. Of course virtually every large city in Eastern Europe has spawned a series of large housing complexes on the periphery and to that extent it could be claimed that the Romanian capital has been endowed with new towns at Balta Alba and Drumul Taberei. It may simply be a question of administrative status.

Then there are several new towns which constitute very large-scale additions to old rural settlements like Bicaz and Onesti (Gh. Gheorghiu-Dej) in Romania, Priboj in Yugoslavia, and Tychy in Poland. (13) The new town of Dimitrovgrad in Bulgaria has involved the coalescence of three village communities and Hungary furnishes several examples of amalgamation: Ajka has developed out of the village of that name, along with the villages of Bode, Tosok, and Tosokberend; Kazincbarcika has grown out of Barcika and Sajokaza; Tatabanya from Alsogalla, Banhida, and Felsogalla, and Varpalota from Inota, Petfurdo, and Varpalota. Other villages with local mining and manufacturing have been annexed by rapidly growing towns in close proximity: thus several mining villages including Diosgyor and Hejocsaba have been attached to Miskolc. (14) Finally there are some new towns which are entirely new developments having no connection with an existing town or village: outstanding here is the case of Eisenhuttenstadt in the GDR and Leninvaros in Hungary. Leninvaros began in 1960 as a chemical combine on the banks of the Tisza producing paints, synthetic resins, and nitrogen fertilizer from natural gas piped from Romania. Polyethylene and olefin plants followed in 1970 and 1975 respectively. Protected from pollution by the prevailing winds in the workers' settlement (Tiszapalkonya) gained urban status in 1966 and in 1970 was named Leninvaros on the centenary of Lenin's birth. Further development has involved an oil refinery and power station. There have been no increases in the gas supply from Romania: instead deliveries of hydrocarbon fuels have been forthcoming from the Soviet Union by pipeline.

These transformations have not been painless. 'Insofar as new towns were determined by natural resources there was practically no choice (of site)' and financial stringency meant that the whole urban structure could not be completed at the same time: 'building began with the residential construction, the water conduit and canalization; later on there followed, as funds became available, the transport, electricity supply, the building of schools, cultural institutions and recreation centres'. (15) Mention should also be made of the structural imbalance of the economy arising from the specific raw material base, creating difficulties for the employment of the female work-force, and delays in integrating new towns into their surroundings as service centres. Social problems tend to arise: 'the preponderance of male youths in the structure of the population goes a long way towards explaining typical characteristics such as breaches of the peace, alcoholism, concealed forms of prostitution etc.' (16) These various problems are generally avoided in new towns which arise as decongestion points. It may not be possible to provide all amenities simultaneously but there is no need to look for functions in regional development since they are related economically to the old centres and commuting may well be involved for employment and various social and cultural needs. Moreover the social structure tends to be similar to that of the old town.

MODERNIZING THE VILLAGES

Some further increase in the number of towns will be necessary, especially in the Balkans where the urban system is still not sufficiently developed to bring all rural areas into a close association. Some commune centres may be able to stimulate interaction with four, five, or six adjacent centres and so justify promotion to urban status. Emphasis on preferential axes may well result in a polycentric system where all commune centres lying on a main line of communication are able to develop in a complementary manner through local specialization. (17) In most cases the system of towns is already well developed and villages are likely to gain urban status only if existing towns and cities extend their territories. But in all parts of Eastern Europe facilities in the villages should be improved.

> The village population is more aware of the relative backwardness of its living conditions than previously, among other things because its composition has undergone radical change: nowadays only 40 per cent of those living in villages are employed in agriculture (thirty years ago this figure was 75 per cent). Thus while the majority of the rural population now work under modern conditions, they live and are housed in primitive circumstances. (18)

Few villages have been comprehensively rebuilt since 1945: some hydro-electric works have necessitated changes and there have been some cases of serious flooding. Also some

238

rebuilding after wartime destruction in Czechoslovakia: Lidice was destroyed as a reprisal for resistance activity while settlements in the Vaclavice area were affected by the German decision to create a large training area with country estates for leading Nazis.

It is necessary to keep enough keen young people in the villages to ensure that the key jobs in agriculture can be taken care of over the longer term. Earnings appear to be broadly satisfactory since the differentials between workers' and farmers' incomes are no longer very great. Yet many young people in the villages today look rather ambivalently on agriculture because substantially longer hours are worked: sometimes as many as ninety hours in agriculture compared with fewer than fifty in industry. Never having owned land (for the collectivization of agriculture was completed more than twenty years ago) they do not have the same 'love of the land' shown by older people and are unwilling to go home 'smelling of the stable'. Better living and working conditions are needed in the villages including good access to the towns and regular vacations. Some farms are already successful. The Podluzan co-operative in south Moravia is based on the old estate of Prisanky where small-holdings allocated under the initial land reform were subsequently amalgamated. The co-operative then joined the neighbouring co-operative of Josefov in 1962 and further mergers have been achieved recently, with Dolni Bojanovice and Mikulcice. This has placed 3,700 ha of agricultural land under a single management. Economies of scale can be made and further investments can be made in farm buildings and welfare services. The 'Red Star' co-operative at Nadudvar in Hungary farms 21.7 thousand ha and is now well known for its computerized management system for handling the livestock enterprises concerned mainly with the rearing and packing of pig and poultry meat. There is also a substantial arable sector involving cereals, sugar beet, and sunflowers. The co-operative has its own retail outlets and is to open shops in Vienna. Housing and community facilities in the village have greatly improved. Particularly outstanding is the community centre with an extensive educational, sporting, and cultural programme.

Non-agricultural employment

Another important factor in the situation is commuting to

non-agricultural jobs outside the village. This has become a commonplace with the fall in farm employment coupled with housing shortages in the towns. It has meant some stability in the population of the villages and some increase in incomes and prosperity but at the expense of long and tiring journeys, almost invariably by public transport. Some commuters travel home only weekly or monthly but then find their family ties diminish. The tendency is for commuters to move permanently to the towns when accommodation is available but many prefer the rural way of life and should be encouraged by better facilities in the villages, because the commuters help to ensure viable rural communities which can attract a full range of services. Better rural living conditions may also be conducive to a modest transfer of town dwellers who move into the country when jobs for skilled workers become available on the co-operatives. The local authorities might then provide a plot of land for a house, or else it may be possible to convert an old barn or take over a house left vacant through the emigration of a German family to the west.

It should not be overlooked that there are a considerable number of opportunities in rural industry; some of these industries are long established. Examples from Hungary relate to mining and quarrying for various substances including ferrous and non-ferrous ores (Recsk, Rudabanya, Urkut, and Uzsa); sugar-beet processing (Acs, Ercsi, Mezohegyes, and Sarkad); ceramics and glass (Herend, Hollohaza, Parad, and Tokad). Other examples could be found in distilling, tanning (including the largest unit in the country, at Simontornya) and wine-making. But in addition it has been appropriate to set up several large new industries in rural areas from the 1960s. New factories have stimulated local housing estates and induced commuting from adjacent villages. Industries include chemicals (Balatonfuzfo, Papkeszi, and Peremarton), metallurgy (Almasfuzito, Apc, and Borsodnadasd), footwear (Bonyhad, Kormend, and Martfu), and food processing (Kaba and Szerencs). Relocation of Budapest industries has contributed to the momentum: engineering at Berettyoujfalu, furniture at Janoshaza, and silk at Szentgotthard for example. Some but by no means all the relocated factories have been set up in the Budapest area. These industries may be attracted by the availability of rural labour and by financial support from the state in the case of location in villages specifically marked out for expansion, because of their accessibility, population size, and range of facilities.

Agricultural co-operatives can also encourage local industries. Since rural crafts attract interest from tourists and city-dwellers, co-operatives now operate central workshops in certain villages and collaborate with master craftsmen who work at home. The trend has been accelerated in Hungary by tax concessions to people engaged in workshops. While in Bulgaria the stimulus comes more directly from the state through the system of agro-industrial complexes. (19) For example Korten in the foothills of Bulgaria's Sredna Gora lies at the centre of an agro-industrial complex covering 4,200 ha devoted to cereals, vineyards, orchards, and stock-raising. Most of the arable land is irrigated with water from the Zhrebchevo dam which passes through the mountains along a 4 kilometre tunnel. The success of the farm's intensive cultures is underlined by the substantial export business in vine seedlings which are sent all over Europe. The industrial component consists of the vinification plant, a repair shop for farm machines, and an engineering factory producing parts for tractor-driven farm machines. With good living and working conditions the migration of young people is not a serious problem.

New factories have to be built in the case of many of these industries but occasionally there are suitable premises already in existence which offer some incentive for local developments: old agricultural machine stations, with teams of skilled workers, have been converted into engineering workshops and redundant mine buildings have been used for textile industries: particularly useful since the latter provide jobs for women in areas of predominantly male employment. Country homes of the former aristocracy and some old military premises have also been converted for rural industrial use. In 1973 exactly a third of Hungary's industrial enterprises (3.38 thousand out of a total of 10.16) were located in the rural areas. Being relatively small (112 workers each on average compared with 171 for urban-based enterprises) and labour intensive (installed power per enterprise 1.64 thousand kW compared with 7.13) their shares for industrial employment and electricity consumption were relatively low at 21.8 and 18.0 per cent respectively but nevertheless the revival of rural industry has been considerable.

THE RURAL SETTLEMENT HIERARCHY: KEY VILLAGES AND OUTLYING HAMLETS

Improvements can be made only on a selective basis because there are too many small villages for all to be viable. It is becoming usual to concentrate investment in the chief settlement in each commune while the outlying villages are progressively depopulated through remoteness from employment and services. A local hierarchy of settlement is often traditional with certain villages providing church services, entertainments, fairs and markets, hostelries, and schools. Selection may be further emphasized by the construction of culture houses and by the reorganization of education of the kind contemplated in Poland, involving considerable amounts of travel, sometimes without adequate transport. Large co-operative farms also play an important role through the construction of administrative buildings and machine stations in the key villages.

In a case study of Baranya, J. Kolta has shown that while the employment in agriculture has fallen from 75.2 per cent in 1949 to 54.9 in 1960, a difference of 20.3 per cent, migration away from the villages amounts to only 11.0 per cent. (20) This shows that a substantial number of people changed their occupation but not their place of residence (Table 7.3). The proportion was much higher in the larger villages and in villages with both bus and rail services. The figures are of course for net migration and take no heed of the various arrivals and departures occasioned by marriage. A hierarchical element is also evident in population trends in Hungary during the 1960s, with the smallest settlements declining most rapidly (Table 7.4) and again in the rural population changes in the mountain section of Romania's Cluj county between 1966 and 1977. (21) The population of 16.17 thousands in the twelve commune centres declined by only 6.1 per cent whereas the figure of 27.18 in the fifty-four other villages fell by 17.1 per cent. Again the population of 26.91 thousands in the thirty villages with connection by national or county roads fell by 11.4 per cent while the population of 16.45 thousands in the thirty-six villages with only local road connections declined more rapidly, by 15.6 per cent. And villages with electricity lost less (15.9 per cent) than those without (17.8).

In the Wismar District of the GDR's Rostock county eighteen central villages (Zentraldorf) were identified for educational, social, and other services and all but five of

Table 7.3 Population and employment in the villages of Baranya, Hungary 1949-60

Population size groups	Employment in agriculture (per cent)			Migration (per cent)
	1949	1960	Difference	1949-60
Below 500	94.3	66.2	-28.1	-20.3
500-1,000	79.9	62.4	-27.5	-12.7
1,001-2,000	74.3	54.4	-19.9	-9.7
Above 2,000	53.8	37.3	-16.5	-4.0
Baranya	75.2	54.9	-20.3	-11.0
Available transport				
Rail and bus	61.8	41.7	-20.1	-5.1
Rail only	72.3	51.2	-21.1	-14.8
Bus only	85.2	59.6	-25.6	-23.7
Neither bus nor rail	83.7	67.4	-16.3	-22.8

Source: J. Kolta (1971) 'The effect of the regrouping of population according to occupation on the functions of rural settlements', in Sarfalvi (ed.) 65-72.

these corresponded with 'church villages', which had previously represented the peak of the rural settlement hierarchy. The five machine tractor stations set up in the district all located themselves in central villages which had emerged out of church villages. In 1960 when collectivization was complete, a plan for agriculture and settlement selected nine main villages (Hauptdorf), all of them being former central villages and each comprising a nodal point on the territory of groups of collective farms (three on average). In 1970 the number of key centres, now labelled 'rural settlement centres', was reduced to seven, taking into account further farm amalgamations, improvements in communications through road surfacing and developments in local organization (especially the formation of village associations dedicated to the improvement of local facilities). All seven had previously been 'central' and 'main' villages and five were part of the historic network of

Table 7.4 Population trends in Hungary by settlement size groups 1960–70

Population size group	Settlements 1970		Population (millions)	Population change 1960–70 (per cent)	Distribution 1960	Distribution 1970
	Number	Per cent				
Below 1,000	1,483	45.0	0.81	-9.2	8.9	7.9
1,001–2,000	831	25.8	1.19	-5.5	12.6	11.5
2,001–5,000	625	19.5	1.91	-2.6	19.8	18.6
5,001–10,000	156	4.8	1.07	2.8	10.5	10.4
10,000–20,000	75	2.3	0.98	6.8	9.2	9.5
20,001–100,000	49	1.5	1.82	14.2	16.0	17.6
Budapest	1	*	1.94	7.5	18.1	18.8

Source: P.A. Compton (1972) 'Internal migration in Hungary 1960–1968', Tijdschrift voor Economische en Sociale Geografie 63: 25–38.

Note: * Less than 0.1

church villages. Thus while depopulation threatens the viability of the outlying settlements a network of key villages maintains a somewhat precarious viability and, despite revolutionary upheavals, there is close continuity with the earlier pattern when the main foci were the villages with churches and estate farms. The appearance of the villages has changed through new farm buildings on the periphery (following the early machine stations which often used the old manor house premises) and new housing in the residential area: 'state-built housing has steadily increased in density from the three and four storey blocks of the early sixties to the high-rise ten storey giants of the seventies which give the skyline of the village a distinctly urban silhouette'. (22) Yet at the same time 'many of the pre-war dwellings have been tastefully renovated and modernised, both internally and externally, and preserve historical continuity at a time of rapid social and economic change'. (23)

Broadly similar measures are being taken in other countries. Regional planning in Bulgaria is being related to individual settlement systems of which some 250 have been recognized. Some have been adequately reorganized while others have been partially restructured or still await attention. The underlying objective is to improve living and working conditions so that each system will become quite a stable entity with only limited migration, thanks to the provision of an adequate range of employments. Romania attaches great importance to the expansion of the urban network by promoting many of the existing villages. (24) Rural communes vary greatly in their potential, according to location and natural resources (Figure 7.2). Several hundred places with relatively good potential were identified in the 1970s and more than a hundred were slated for promotion by 1980 (Figure 7.3). The planners concentrated their attention on communes which already had significant local industries and a history of commercial activity (in connection with former administrative and marketing systems), especially communes remote from existing towns. The programme has fallen behind schedule and the promotions due by 1980 have not yet taken place. But building work continues and the leadership has reasserted its commitment to rural planning, partly to mobilize the work-force more efficiently for non-agricultural employment and partly to create more land for agriculture through the eventual clearance of all

245

Figure 7.2 The development potential of Romania's communes

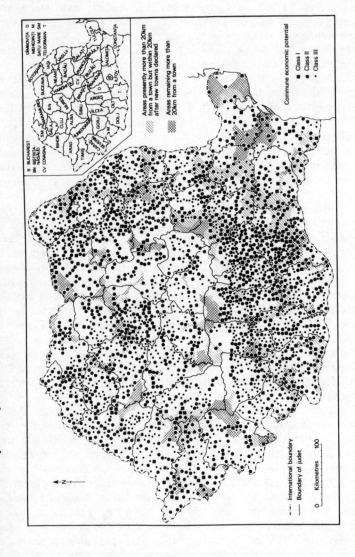

Source: Field-work.

Figure 7.3 Romania's new town programme

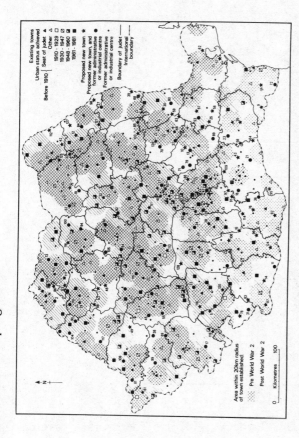

Source: D. Turnock (1986) 'The rural development programme in Romania with particular reference to the designation of new towns', Leicester University Geography Department Occasional Paper 13. See pp. 9–10 of this work for the names of the new towns shown as serial numbers 1–129.

Figure 7.4 Planning in the Cluj region, Romania

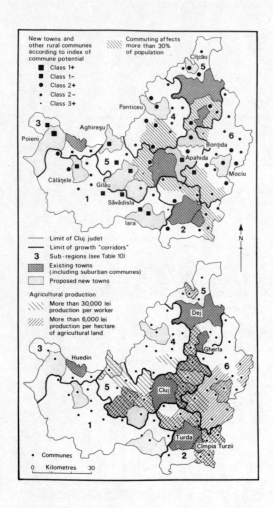

Source: D. Turnock (1986) 'The rural development programme in Romania with particular reference to the designation of new towns', Leicester University Geography Department Occasional Paper 13.

Notes: Key to subregions: 1. Apuseni Mountains 2. Aries Valley 3. Cris Valley 4. Somes Plateau 5. Somes Valley 6. Transylvanian Plateau

248

settlements lying outside the designated building perimeters. There is also a desire to control the expansion of large cities, so in the Cluj region of Transylvania there are proposed new towns not only in the remoter districts (for example Iara, Mociu, Panticeu, and Poieni) but also in places close to the city (Apahida, Bontida, Gilau, and Savadisla) where accommodation for commuters will be provided. Over the long term, development corridors will extend from Cluj-Napoca to towns of Huedin in the west, Gherla and Dej in the north, and Turda and Cimpia Turzii in the south (Figure 7.4).

Hungary has a 'National Settlement System Development Conception' which aims at better professional services, shops, roads, and power supply. But problems arise because of the lack of harmony between economic plans and settlement plans: co-operative and state farms hold together several villages economically but this reality is not reflected adequately in physical planning because farm offices may not be situated in the key settlement and new housing for employees may be irrationally located. (25) Much depends on the quality of local leadership if modernization is to be as rapid as possible. A study of villages in the Kecskemet area revealed several cases where much-needed development had been frustrated by local authority support for sectional interests. It was stated that the consumers' co-operative in Kunszentmiklos paralysed the endeavour of a retailing company to build a department store fearing that its position would be weakened in the process. However the department store was very necessary, especially since the local co-operative had shown itself unable to establish an adequate network of shops.

The outlying settlements

Rural life has been altered irrevocably in the post-war years. The reorganization of agriculture has been a force for modernization for in the wake of the state farms and co-operatives has come electricity, piped water, and new roads. Radio and television now occupy rural dwellers during their leisure time to a greater extent than traditional preoccupations, to say nothing of the increase in organized culture and travel. But rural development is a slow process and it is predicted that even at the end of the century only half the communes will possess a fully equipped centre. In the remotest areas the majority of the population may still depend on relatively low agricultural incomes, with non-agricultural employment restricted to occasional opportunities in building and construction. Furthermore public transport and local distribution may be so poor as to lead to much frustration and wasted time simply to secure basic items like bottles of gas. So the younger people will inevitably leave. The authorities have adopted an ambivalent attitude. They cannot invest heavily in the smaller outlying settlements and have tended to amalgamate small units for administrative convenience. But draconian measures to force resettlement, perhaps by restrictions on building and repairs, have not been imposed.

In Czechoslovakia there has been a great reduction in the number of villages but only as a result of administrative reorganization. There were 15.25 thousand villages in 1930 but 14.80 in 1949, 10.73 in 1964, and 8.88 in 1977. Some settlements were effectively abandoned as a result of the expulsion of Germans after the Second World War but most of the changes have been motivated by the desire to focus on viable rural settlements with services, rural industries, and a co-operative farm headquarters (including offices, garages, workshops, stores, silos, and fodder units). However, although no longer recognized as separate settlements administratively, many small outlying hamlets survive with relatively backward conditions which the official statistics do not expose. The elimination of differences between town and country inevitably requires the eventual disappearance of the remoter places. So if people decide to build a house in a non-central village they must realize that within one or two decades the local school and shop will probably disappear. The future is to be sought in villages with some 2,000 inhabitants. However, mobile shops provided for sparsely populated areas like the Sumava, run by the local consumer co-operative and medical services in the countryside, have been improved with the establishment of a network of branch polyclinics along with mobile units dealing with prophylactic examinations and dental services.

Romania's rural plan involves the eventual abandonment of half the country's 14,000 villages. These are places which are considered to have no potential for development because they are small and remote, with few resources and services. It is likely that people will choose to leave the remoter places and settle in or close to towns where non-agricultural employment is available but the coercive element in the Romanian strategy is remarkable. New building can take place only within designated building perimeters and these are laid down only in villages considered to have potential for development (although even in such villages the official perimeter is usually smaller than the present built-up area, thereby inducing consolidation by reducing courtyards and constructing more two-storey houses). In the Nehoiu and Patirlagele areas of the Buzau region shown in Figure 7.5 building perimeters are smaller than the present settlement area in virtually every case. Settlements without perimeters will eventually disappear. Already it is difficult to carry out structural repairs in such places and where the land is considered suitable for arable farming there are many cases of compulsory evacuation. The destruction of villages in Transylvania in 1988 was particularly controversial and caused a crisis in relations between Romania and Hungary where the rural planning was interpreted as discrimination against members of the Hungarian minority. The issue is controversial not only because it impinges on individual liberty but also because it raises questions about the assessment of potential for individual settlement. Remote villages could have considerable potential for tourism and given better roads and greater car ownership in future the outlying villages could become attractive commuter settlements. These issues are discussed further in the final section of the chapter.

Hungary's tanyak

The situation is particularly difficult in areas of Hungary where detached farms (tanyarendszer) predominate. (26) Some 200,000 households are involved. Housing conditions are poor, for the average number of rooms per dwelling is low and many of them have walls of adobe, mud, or beaten earth built before the Second World War. A high proportion lack electricity and public transport while cultural and educational opportunities are poor. Quite apart from the tanya problem there are some 700 nucleated settlements with populations below 500. The total population involved is some 750,000, almost 7 per cent of the total population of the country. But depopulation is rapid: industry is totally absent, the institutional network is inadequate and the creation of an infrastructure is uneconomic. Lying at the

Figure 7.5 Settlement and rural planning in the Nehoiu and Patirlagele areas of the Buzau region, Romania

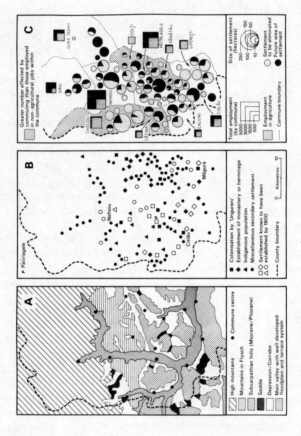

Source: D. Turnock (1986) 'The rural development programme in Romania with particular reference to the designation of new towns', Leicester University Geography Department Occasional Paper 13.

bottom of the central place hierarchy these dwarf villages are considered officially to have no potential for growth and morale has declined as the younger and more active people leave for the cities. The problem is compounded by the fact that these problem villages are clustered in the more backward parts of the country where depopulation is general and where villages which once had substantial populations are falling into the category of basic settlement units with only residential functions. Hamlet aggregations are most noticeable in the hilly regions with fragmented terrain and poor communications. The average population of each village is only about half the national average of 1,700 in the counties of Baranya, Vas, Veszprem, and Zala, all in Transdanubia. On the other hand the villages are relatively closely spaced with approximately seven villages per hundred square kilometres compared with a national average of 3.4.

There have been attempts made since 1948 to regroup the population of the tanyak in nucleated settlements but it is difficult to do this without a substitute for the intensive garden culture of the Kecskemet area: co operative farm workers may reside on their isolated farms and grow fruit and vines on the small-holdings which comprise about half a hectare of land around the house. The tanyak can have some advantages for the co-operatives because even though they may interdigitate with the communally farmed land the work-force is dispersed and transport to the fields is simplified. Arguably more effort should be made to support the present settlement pattern in contrast to the centralizing tendencies encouraged by the 1971 settlement concept. Depopulation accelerated during the decade and Gyurufu in Baranya country became totally deserted. However some revisions to the plan have been made since the local authorities most concerned with the tanya problem (Baranya, Borsod, Gyor-Sopron, Somogy, Vas, Veszprem, and Zala) convened a conference in 1978 in the Zala village of Lenti to review the situation. Despite the provision of mobile cultural and medical facilities for small settlements there is general expectation that many individual farmsteads and small villages will disappear by the end of the century.

RURAL PLANNING: THE SCOPE FOR RECREATION AND TOURISM

In Czechoslovakia the number of second homes has grown since the 1930s to reach 40,000 in 1945 and some 220,000 in 1985. Two-thirds of all Prague families now have their own cottage (or access to somebody else's) or the use of enterprise recreation facilities. Many old farm houses have been taken over as the agricultural population builds new accommodation. Much the same situation has been reported in the Brno area. Construction rates have increased from some fifty per annum in the 1930s to 200 in the late 1950s and more than 2,500 by 1970, with sites near reservoirs and suitable for gardening being particularly popular. A boom in second-home ownership occurred during the 1970s with improved living standards: a shorter working week, higher incomes (through improved wage rates and greater opportunities for moonlighting) and rising levels of car ownership (only three cars per thousand of the Hungarian population in 1960 rising to 100 in 1981). People wish to use their leisure time constructively and restrictions on foreign travel have led many families to concentrate their resources on securing a second home. Even in the Balkans there is a growing emphasis on weekend homes, many of them built illegally on farm land and many of them unnecessarily large as well as wasteful of building materials and manpower.

The second-home movement obviously poses problems of an ethical nature: how can the state contemplate the occupation of two homes by certain families when many others are restricted to a single apartment in squalid and overcrowded conditions? However, the more affluent families who enjoy such facilities are usually closely connected with the party and it would seem counterproductive for the authorities to infringe property rights in such cases. Moreover not all second homes involve cases of ownership of two houses and in many cases the second home may provide a safety valve for a family confined to a small rented urban apartment during the week: the older members of a family may take up permanent residence in a second home; this could reduce overcrowding in the urban apartment or even free an apartment altogether. In this way the state is relieved of some of the pressure that might otherwise exist for greater investment in housing and recreation facilities. So it is even possible for families to obtain loans from the national savings bank to finance the purchase of a second home. It is also relevant to consider local politics because sales and leases of building plots can boost the income of local authorities and so contribute to the decentralization of decision-making which is one of the principles behind the New Economic Mechanism. The toleration of second homes, once available only to a small privileged minority, is a reflection of the more pragmatic and less doctrinaire socialism which would appear to be a suitable vehicle for economic and social progress in Eastern Europe.

The second home movement needs careful planning however. Research in Hungary shows that second homes are most numerous in the scenic areas close to the major cities. In Hungary almost three-quarters of the 127,000 second homes in existence in 1980 were on Lake Balaton, Lake Valence, the Danube Bend, and the immediate environs of Budapest. Another 7 per cent were located in the Bukk, Matra, and Villany hills. At Balaton the standards of construction and the quality of the infrastructure is relatively high. And elsewhere there are some developments which have been orderly in environmental and landscape terms, for example at Szanazung near Bekescsaba in the southern section of the Great Plain where roads, water, and drainage have been installed prior to house-building on woodland plots. However in many cases, and especially around Budapest, development has not been controlled and many cabins exist illegally without proper facilities. Plots of land allocated as gardens at the time of collectivization are naturally endowed with sheds which, given a scenic setting, may easily become weekend houses. Second homes can contribute to village life by providing business for local shops and restaurants but the high levels of concentration in certain settlements combined with intermittent occupancy can have negative implications for village society. In several cases more than a third of all houses may be weekend cottages and in some instances the proportion may approach two-thirds as in the villages of Balatonudvari near Veszprem and Zamardi near Siofok with values of 56 and 64 per cent respectively.

The resort village concept has gained momentum during the 1980s with house sales to individuals and also to large companies who can make the properties available to their employees. In Bulgaria it has been suggested that the most efficient way to provide additional recreational facilities for city-dwellers would be to rebuild or restore old and abandoned village houses: a new state enterprise might locate, purchase, restore, and finally sell old village houses. In some regions abandoned houses account for more than three-quarters of the total stock (more than 90 per cent in some localities) and over the country as a whole a thousand villages may disappear from the map by the end of the century. Since holiday accommodation currently provided by the state and the various economic enterprises is inadequate there is good reason why the authorities should look favourably on such a scheme of restoration. It is now possible that many of the thousands of empty houses in dwarf villages will be put to good use especially when they fall within the thirty resort zones which have been identified, lying outside the developing regions where there is a continuing housing shortage which rules out house sales for vacation use. So collective farms have also been active in selling or leasing land for second homes. Such practices help to reduce unauthorized development thereby facilitating the controlled evolution of the landscape. Outstanding here has been the development at Balatonudvari where three sites have been developed since the late 1960s. The whole scheme is environmentally sound with a regular ground plan and an orderly development process covering housing and infrastructure, while financially it has been profitable to the local authorities.

There is also scope in a wide development of tourism with improvements to farm houses to provide accommodation for visitors. In Yugoslavia money earned through work in Western Europe can go into tourism, especially in view of the restrictions on the size of private farms. Local authority grants are also available to improve houses in Slovenia and the rural developments help to support the revival of inland resorts like the spa of Rogaska Slatina and the ports of Koper and Piran. (27) Elsewhere in Eastern Europe it would appear that many new jobs could be provided in rural areas through tourism and it would appear that work in agriculture and tourism could be more satisfying than jobs on the factory production lines. A shoe factory at Novy Targ in Poland was appreciated in the 1950s as offering some relief to heavy population pressure on the land but the loss of peasant independence has led many of the employees to leave, especially on account of unsympathetic management attitudes after a strike in 1976, and the factory has been reported running below capacity because of the labour shortage.

Tourism and agriculture in the mountains of Eastern Europe

Some of the best prospects for tourism lie in the mountains where there are outstanding scenic attractions, opportunities for winter sports and mineral water treatment, and a rich cultural heritage based on a relatively strong element of continuity in settlement patterns and economic structures. The marginal nature of land has discouraged state investment in agriculture while the mineral resources have generated some powerful growth centres which have become poles of attraction in the migration process. Tourism could now be a means of stemming the exodus from the farming districts. There are however some major differences in approach when the policies of one government are compared with another. In Albania an uncompromising autarkic policy demands the fullest use of mountain land: intensification is evident on the planting of orchards and the extension of maize cultivation right up to the physical limits of soil and climate. But in Slovenia migrant workers returning from a sojourn in the west are investing in new houses in mountain villages with back-up from government spending on electrification, roads, water supplies, and even small industries. Tourism helps to justify modernization of the mountain settlements on this scale although it may lead to some neglect of agriculture; but arguably a comprehensive advance by tourism will be accomplished most readily where private as well as state investment can be envisaged. This requires a permissive environment where the state tolerates and encourages individual initiative and a land-holding system which enables individuals to capitalize on their assets. There is a large private agricultural sector in Poland and Yugoslavia but elsewhere it is small, comprising a residue of farms not readily amenable to co-operative management. However such farms may be very important in certain localities, typically mountain districts with highly dispersed settlement patterns offering many advantages for tourism.

The Roman Carpathians

With a total area of 1.41 million hectares private farms account for less than 10 per cent of the total agricultural land in Romania (Table 7.5). Yet in a number of

Table 7.5 Individual farms in Romania

(a) Landholding and Livestock 1970–80 by regions

| Region | Share of total land/stock belonging to individual farms (per cent) | | | | | | | | | | | |
| | Agricultural land | | Arable land | | Cattle | | Sheep | | Pigs | | Poultry | |
	1970	1980	1970	1980	1970	1980	1970	1980	1970	1980	1970	1980
Centre	21.1	21.4	14.3	15.2	25.4	28.2	30.9	30.4	23.6	20.9	na	15.5
North	13.9	14.0	10.2	10.4	20.0	19.7	19.2	19.8	16.3	19.2	na	13.1
North-east	7.8	8.0	2.6	2.7	13.6	14.8	9.4	9.1	6.6	6.1	na	8.8
South-east	3.9	4.2	1.2	1.4	8.6	9.1	6.4	8.7	5.0	4.3	na	8.8
South-west	9.5	10.1	5.2	5.6	14.4	16.3	13.2	13.0	7.0	6.5	na	8.5
West	9.7	9.6	7.4	7.4	15.9	16.0	14.5	15.7	6.7	7.7	na	11.0
Romania	9.2	9.4	4.6	4.8	15.0	15.9	12.7	13.8	8.2	7.9	na	8.8

(b) Production of main crops by farm type 1960–80

Crop		State farms			Percentage share falling to Co-operatives			Individual farms		
		1960	1970	1980	1960	1970	1980	1960	1970	1980
Cereals	Area	13.4	16.6	18.3	65.3	68.6	68.7	19.5	5.5	5.0
	Output	17.1	24.1	22.0	66.2	63.5	64.1	14.1	3.4	3.9
Oil plants	Area	15.1	27.3	25.3	75.6	72.6	74.7	9.2	*	*
	Output	14.3	33.6	28.8	75.7	66.4	71.1	9.7	*	*
Sugar beet	Area	2.8	0.1	0.1	95.0	99.9	99.9	2.2	*	*
	Output	2.7	0.2	0.1	95.5	99.8	99.8	1.7	*	*
Potatoes	Area	3.7	5.0	4.3	41.7	41.6	39.9	45.3	20.2	22.7
	Output	3.7	7.9	5.7	44.3	42.2	35.9	43.0	15.4	20.3

Source: Statistical yearbooks.

Note: * Less than 0.1

administrative regions in the Carpathians the proportion rises almost to a quarter. (28) There are some quite large areas where co-operative working was considered non-viable when the programme of collectivization was brought to a halt in the early 1960s. However agriculture on the small farms is no longer as intensive as it used to be. It seems that a peak was reached in the late nineteenth century with a combination of cereal cultivation and livestock extending to outlying grazings as well as the lands immediately adjacent to the village. (29) Some reduction of arable farming in unpromising situations accompanied a rising tempo of internal trade following the construction of railways. But the trend has been maintained, and it may be that the retreat from poorer land has accelerated since 1948 through compulsory delivery quotas and through land taxes which seem to be greater in relation to the value of production than before the installation of a communist regime: taxes relate to buildings and farmland with the latter divided into different categories according to land use while subjectivity in assessment may involve conscious discrimination which is bound to affect initiative.

Individual farms in the Romanian Carpathians face a difficult situation. Although there is a degree of independence which the farmers appreciate, the small scale of the operation (with regular employment of non-family labour forbidden) limits the scope and it is evident that with redistribution of population in Romanian Carpathians over more than half a century the main areas of individual farming (the Western Carpathians) have suffered most acutely from depopulation (Figure 7.6). If anything, farms appear to be decreasing in size owing to the traditional form of inheritance which permits subdivision of the family farm: when the eldest son marries he can claim his portion of land and on the death of the father all the land is subject to reallocation with sons and daughters treated on equal terms. In practice however the amount of subdivision and further fragmentation of holdings is often quite modest because many sons and daughters leave their home village and may wish to sell their interests. The attractiveness of a mountain farm has been further reduced by the hiving off of woodlands. This reduces the wood supply available to small processing industries and eliminates ancillary employment that could once be generated within the community. Loss of woodlands is complemented by a prohibition on the use of firearms by peasants (and indeed of any form of hunting in forests), for the sport is a lucrative industry for the state and a tendency for the stock of wild animals to increase has adverse implications for sheep farming on outlying lands. Moreover traditional practices of grazing livestock in young plantations, with suitable protection for the trees, are no longer tolerated. At present the rural planning programme has worrying implications for private farmers who will be left a long way from their lands if pressure is exerted on families living in isolated farmsteads to move into restricted building perimeters in the principal villages: at present some compromises appear to be being worked out which will allow the traditionally dispersed settlement pattern to be perpetuated officially.

Much seems to depend on the availability of non-agricultural employment in close proximity to the farm. Where such work is available then private farms are considered most acceptable by young families because they provide (to a greater extent than the private plots of co-operative farm members) food for the household and some additional income through sales on local private markets and by contracts with the state. In one village within range of Brasov the importance of work in industry is now very great:

> the peasant domestic group can better fulfil expectations of consumerism by adding a shift rather than a cow: agricultural expansion in comparison to industrial participation yields less per unit of labour while the industrial worker gains valuable medical and social services, not the least of which is social security in old age, which are not available through private agricultural work. (30)

Many youthful families therefore take the fullest opportunities for work outside the village. Commitment to agriculture is reduced with a run-down in arable and concentration on dairy cattle which can harmonize with factory shift work. The place of residence may be moved from a position on the hillside, where the best potato ground may lie, to a new site in the valley convenient for commuting to work: here a small vegetable plot and garden may be cut out of a steep hill slope despite the danger of erosion. A small house may be retained on the higher ground as a base for either agriculture or recreation. And in the same way country houses with a little land may be valued by families who have moved right out of the village and taken

Figure 7.6 Redistribution of population in the Romanian Carpathians 1911-77

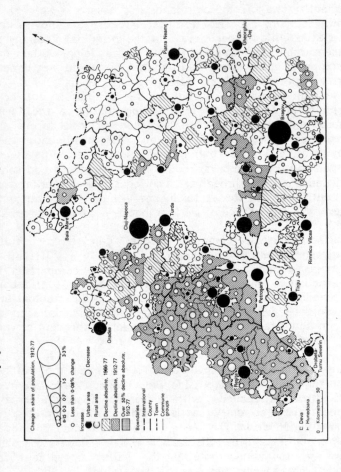

Source: Census.

an apartment in town. Yet some young families choose not to participate in the industrial sector at all and they stay higher up the hill, living alongside elderly relatives who can be of great assistance in maintaining a significant agricultural effort.

Even so the farming is not as intensive as it used to be. Evidence from the Carpathian village of Corbu near Borsec, studied by British geographers in both the 1930s and 1970s, shows the trend towards simpler cropping patterns. (31) Several crops grown in the 1930s, including lucerne, hemp, oats, and rye, are not grown today while the only new crops are poppies intersown with part of the potato crop and tree fruits (apples and plums). Otherwise the crops remain the same (maize, potatoes, vegetables, and grass), but the proportions have changed: the area of maize has been considerably reduced and in its place the potato and grassland areas have expanded. Cropping is discouraged by the difficulties of mechanization. There are no individually owned tractors and where hiring from local machine stations is possible not all the land may be accessible or otherwise suitable for working. On the other hand in remote villages where daily commuting to work is not feasible there are very clear signs of depopulation and reduction in agricultural effort. Young people move away and farms are abandoned when the parents die. Work on the remaining farms is concentrated in the hands of women while the menfolk are away in the towns throughout the week. Recent Romanian research suggests that only one school-leaver in eight in private farming communes chooses to stay at home, with the result that the agricultural population not only declines but also becomes more aged (70 per cent of private farmers being over 40 years of age). The local infrastructure is frequently poor: while houses have often been modernized cowsheds remain cramped and unhygienic and when transport is poor (only steep footpaths to give access to extensive grazing areas on the high ground) it is difficult to ensure that land is adequately fertilized. Inaccessibility also inhibits the growth of farm-based tourism, although further constraints arise from the state's emphasis on organized parties travelling by coach and the prohibition on the use of private accommodation for foreign tourists.

Very considerable changes are evident in the extent of stocking high pastures. Remote grazings are often under-used (while land close to the farms may be over-grazed) and further fodder is lost through lack of integration between farming and forestry. Less use is made of the distant pastures where hard work was formerly encouraged by integration with cultural manifestations that lay right at the core of community life. Transhumance movements have therefore become more local in character, involving either sedentary shepherding on the day pastures or pendular transhumance on a seasonal basis between village and 'stina'. It is common for the pastures in which the stina is situated to be owned by the village community some of whose members will work in the mountains for the season (usually May to September). As well as the local flocks, animals from other villages may be taken over. In such a situation the local authority, in the form of the communal People's Council, plays an important role in the organization which has become more important in the post-war period with the elimination of the system whereby a middleman would rent a stina and employ a head shepherd (baciu). The local council may also own land close to the centre of the village and rent it out to farmers who have insufficient land of their own for cropping or grazing. However there is still some individual enterprise involved in the stina system because a shepherd and his family may take over sheep from other farmers and look after them for the season in return for a cash payment plus a share of the cheese and milk produced: hence the ritual testing of the productivity of each animal as part of the handing-over ceremony. Sheep may also be taken from low-ground co-operatives on similar terms, although it is also possible for the latter to maintain their own summer grazing stations.

A typical stina is used by about five hundred sheep, a few young cattle, and half a dozen pigs, the livestock of some seventy families, attended by ten men and boys (women may appear, but it is not customary). Work is organized by the head shepherd, an occupation usually handed down from father to son, though much of it conforms to a well-established routine. Sheep are milked three times a day and most of the time is taken up with cheese-making. This is the main produce from the stina and is stored pending transfer to the village from time to time. The stina displays few signs of modernization: apart from the intrusion of enamel and aluminium tableware among traditional wooden utensils the most obvious concessions to the 1980s are plastic sheets (to repair leaking roofs) and transistor radios. In the case of cows' milk produced close to or within the village there is now greater scope for marketing in liquid form, either to nearby towns or else to milk-processing plants where cheese, milk powder, and other products are made. All milking, whether on the farm or the stina, is still done by hand as farm incomes are just too small to allow mechanization even supposing that suitable equipment was available for purchase.

What of the future of the private farm? The private farmer values his independence and a radical transformation through compulsory collectivization is not likely to be pursued any more in the future than it was in the past. Equally however the value of independence becomes economically weaker as farm incomes fail to keep pace with improvements in material prosperity in the country generally and socially less compelling as depopulation erodes local community spirit. The logical solution of allowing larger and more efficient private farms to emerge, perhaps with government grants for improvements to roads and buildings, is ruled out by the ideological emphasis on state and co-operative ownership of the means of the production. A pragmatic approach demands some special arrangements but embarrassing political implications must clearly be minimized. It is easy to propose improved education, to provide young people with qualifications relevant to progressive mountain agriculture, better livestock breeding facilities (artificial insemination stations) and veterinary services, as well as better arrangements for fodder, perhaps through a chain of agricultural stores. Light machinery should be made available, tourism encouraged, and incentives offered whereby cottages can be built or renovated for tourist use. There could also be further decentralization of food-processing industries. But the crucial factor of course is the farm structure and here the signs are that the state favours a compromise between the private farm and conventional co-operative through small associations, which could evolve through various stages by simple co-ordination of grazing, fodder storage, and marketing followed by construction of buildings, purchase of machinery, and integration with forestry. Small associations could also include local industries in their activities. Such structural changes would satisfy the state's insistence on 'socialist' institutions and might be flexible enough to avoid complete concentration of initiative in the hands of the authorities. The formula would also be conducive to some consolidation of dispersed settlement and harmonize with

Figure 7.7 Development level of communes in the Romanian Carpathians

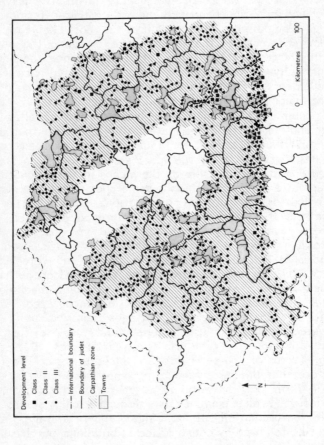

Source: D. Turnock (1986) 'The rural development programme in Romania with particular reference to the designation of new towns', Leicester University Geography Department Occasional Paper 13.

the ideals of the rural planner for compact settlement groupings which can be properly serviced. (32)

Surveys of the past performance of mountain communes suggest that the growth potential is very variable (Figure 7.7) so there could be considerable advantage in encouraging more local initiatives, enabling redundant farmhouses to be used as tourist cottages under the management of active farmers in the community. At the moment however neither the state nor the local authorities are interested in switching the emphasis away from places where large-scale developments will eventually justify promotion to urban status. This is in spite of the fact that the evacuation proposals for 'non-viable' settlements have caused great resentment and have been a factor in the recent migration of Romanian citizens (including many ethnic Hungarians) from Transylvania across the frontier to Hungary, where demonstrations have focused specifically on the destruction of the countryside. Thus it is in the rural areas (the residual areas of socialist modernization) that the different roads to socialism in Eastern Europe are particularly apparent.

NOTES

1.   Halpern (1972); Hann (1980); Simic (1973); Verdery (1983). See also A. Simic (1973) 'Kinship reciprocity and rural-urban migration in Serbia', Urban Anthropology 2: 205-13; A. Simic (1973) 'The best of two worlds: Serbian peasants in the city', in G.M. Forster and R.V. Kemper (eds) Anthropologists in Cities, Boston, Mass.: Little, 179-200.

2.   Mihailovic (1972) 83.

3.   S. Herman (1977) 'Urbanization processes in Poland 1960-1970', in Kuklinski (ed.) 105-50.

4.   Z. Tatai (1976) 'The growth of rural industry', in Enyedi (ed.) 71-88. See also G. Krajko and R. Mrszaros (1980) 'Effects of industrialization on economic and social change in village areas on the south Hungarian Plain', in D.R. Deskins (ed.) Impact of Urbanization and Industrialization on the Landscape, Michigan Geographical Publications 25: 166-94.

5.   Z. Tatai (1976) 'The growth of rural industry', in Enyedi (ed.) 71-88.

6.   Mihailovic (1972) 90.

7.   ibid. 91.

8.   H. Adamczewska-Wejchert and K. Wejchert (1975)

'Spatial structure of middle-sized towns', Geographia Polonica 32: 93-104.

9. I. Ianos (1984) 'Geographic considerations on the commercial activities of Romanian towns', Revue roumaine: geographie 28:49-52; D. Turnock (1987) 'Urban development and urban geography in Romania: the contribution of Vintila Mihailescu', GeoJournal 14: 181-202.

10. S. Herman (1977) 'Urbanization processes in Poland 1960-1970', in Kuklinski (ed.) 105-50; A. Jedrasko (1977) 'The migration of population and its influence on planning of urban agglomerations in Poland', in Kuklinski (ed.) 84-104; Z. Rybicki (1977) 'Territorial structure of Poland', in Kuklinski (ed.) 7-36. See also W.C. Biselle (1971) The Polish Peasants: Change and Continuity in the Post-War Era, Syracuse University: Ph.D. thesis; W.C. Biselle (1983) 'The peasant worker in Poland', Studies in European Society 1: 26-39; M. Dobrowolska and J. Herma (1968) 'Migration of manpower in south Poland as a factor of change in regional structure', Geographia Polonica 14: 321-31; M. Dobrowolska (1978) 'Changes in agrarian structure of the villages of southern Poland', Geographia Polonica 38: 41-8; A. Gawryszewski (1981) 'Spatial mobility of the population of Poland: an attempt of an integrated approach', Geographia Polonica 44: 225-38; Hann (1985); G. Kolankiewicz (1980) 'The new 'awkward class': the peasant worker in Poland', Sociologia Ruralis 20: 28-43.

11. L. Vriser (1984) 'Yugoslav national settlement system', in Bourne et al. (eds) 400-11.

12. H. Grimm (1984) 'The settlement system of the GDR: its structure and development', in Bourne et al. (eds) 377-99.

13. J. Wawrzynski (1986) 'New towns concept in Poland', in I. Masser and R. Williams (eds) Learning from Other Countries: The Cross-National Dimension in Urban Policy-Making, Norwich: Geo Books.

14. F. Boros (1970) 'Geographical aspects of Dunaujvaros', in Sarfalvi (ed.) 55-64; F.W. Carter (1975) 'Bulgaria's new towns', Geography 60: 133-6.

15. Mihailovic (1972) 96.

16. ibid. 97.

17. A. Zagozdzon (1972) 'Semi-urbanisation with special reference to the problem of settlement complexes', Geographia Polonica 24: 57-69.

18. L. Lacko (1976) 'The hamlet and farmstead regions of Hungary: planning problems', in Enyedi (ed.) 59-70

(quote p. 66).

19. G. Barta (1976) 'Changes in the living conditions of the rural population', in Enyedi (ed.) 89-110; I. Volgyes (1980) 'Economic aspects of rural transformation in Eastern Europe', in Volgyes et al. (eds) 89-127.

20. J. Kolta (1970) 'The effect of the regrouping of population according to occupation on the functions of rural settlements', in Sarfalvi (ed.) 65-72.

21. P. Bell (1979) Social Change and Social Perception in a Rural Hungarian Village, University of California at San Diego: Ph.D. thesis; P. Beluszky (1976) 'Functional types of rural settlement in Hungary', in Enyedi (ed.) 41-58; P.A. Compton (1972) 'Internal migration in Hungary 1960-1968', Tijdschrift voor Economische en Sociale Geografie 63: 25-38; A. Gyenes (1973) 'The restratification of the agricultural population in Hungary', Acta Oeconomica 11: 33-49; P. Halmos (ed.) (1972) 'Hungarian sociological studies', University of Keele Sociological Review Monograph 17; M. Hegedus (1974) 'Some factors influencing urban developments in Hungary', Acta Oeconomica 12: 171-89; N. Horchner (1983) 'Urban planning in Hungary', Planning Administration 10: 71-6; K. Kulcsar (1983) 'Rural development in post-war Hungary', Sociologia Ruralis 23: 204-10; M. Vorosmarti and E. Tajti (1976) 'Demographic characteristics of rural settlements', in Enyedi (ed.) 27-40; G. Zovanyi (1986) 'Structural change in a system of urban places: the twentieth century evolution of Hungary's urban settlement network', Regional Studies 20: 47-71.

22. V.D. Freeman (1979) 'Agricultural development and rural change in the GDR', Sheffield City Polytechnic Department of Geography and Environmental Studies Occasional Paper 1, 79.

23. ibid.

24. C. Lazarescu (1976) 'Current problems of organizing the national territory and human settlements in Romania', Revue roumaine: sciences sociales 20: 37-44; S. Sampson (1976) 'Feldioara: the city comes to the peasant', Dialectical Anthropology 1: 321-47; S. Sampson (1982) The Planners and the Peasants: An Anthropological Study of Urban Development in Romania, Esbjerg: University Centre of South Jutland Institute of East-West Studies; S. Sampson (1984) 'Elites and mobilization in Romanian villages', Sociologia Ruralis 24: 29-51; Sampson (1984); D. Turnock (1976) 'Restructuring of rural settlement in Rumania', Slavonic and East European Review 54: 83-102; D. Turnock

(1986) 'The rural development programme in Romania with particular reference to the designation of new towns', Leicester University Geography Department Occasional Paper 13; D. Turnock (1987) 'Romania', in Dawson (ed.) 229-74; I. Velcea (1972) 'The urbanisation process of rural settlement in Romania', Revue roumaine: geographie 16: 93-101.

25. G. Enyedi (1981) 'The effect of modern agriculture upon the development of the rural settlement network in Hungary', in C. Csaki et al., Agriculture and Regional Development in Europe, Budapest: Research Institute for Agricultural Economics.

26. E. Lettrich (1969) 'Hungarian tanya settlement: history and present day problems', in Sarfalvi (ed.) 151-68; E. Petri (1969) 'Collectivization of agriculture and the tanya system', in Sarfalvi (ed.) 169-80; N. Volgyes (1980) 'The Hungarian tanyas: persistence of an anachronistic settlement and production form', in Volgyes et al. (eds) 175-90.

27. A. Barbic (1983) 'The farm worker in Yugoslavia: a bridge between the city and the country', Sociologia Ruralis 23: 76-84; F.W. Carter and R.A. French (1975) 'New era in Slovenia', Geographical Magazine 47: 556-60; W. Lockwood (1973) 'The peasant worker in Yugoslavia', Studies in European Society 1: 91-110; I.M. Matley (1968) 'Transhumance in Bosnia and North Hercegovina', Geographical Review 58: 231-61; C. Thomas (1974) 'Rural-urban interaction in western Slovenia', in Kosinski (ed.) 245-65; C. Thomas (1978) 'Decay and development in Mediterranean Yugoslavia', Geography 63: 179-87; C. Thomas and M. Vojvoda (1973) 'Alpine communities in transition: Bohinj, Yugoslavia', Geography 58: 217-26.

28. J.W. Cole (ed.) (1984) 'Economy society and culture in contemporary Romania', Department of Anthropology University of Massachusetts Amherst Department of Anthropology Research Report 24; D.A. Kideckel (1976) 'The social organization of production on a Romanian co-operative farm', Dialectical Anthropology 1: 267-76; D.A. Kideckel (1977) 'The dialectic of rural development: co-operative farm goals and family strategies in a Romanian commune', Journal of Rural Co-operation 5: 43-62; D.A. Kideckel (1982) 'The socialist transformation of agriculture in a Romanian commune 1945-1962', American Ethnologist 9: 320-40.

29. E. Lichtenberger (1978) 'The crisis of rural

settlement and farming in the high mountain region of continental Europe', Geographia Polonica 38: 181-7; I.M. Matley (1970) 'Traditional peasant life in Romania', Professional Geographer 22: 311-6; I.M. Matley (1971) 'Human geography in the Western Mountains', Scottish Geographical Magazine 87: 116-27.

30.   S.G. Randall (1976) 'The family estate in an upland Carpathian village', Dialectical Anthropology 1: 277-85 (quote p. 281); S.G. Randall (1982) The Household Estate under Socialism, University of Massachusetts, Amherst: Ph.D. thesis. See also S. Beck (1976) 'The emergence of the peasant-worker in a Transylvanian mountain community', Dialectical Anthropology 1: 365-75.

31.   Turnock (1980).

32.   D. Turnock (1986) 'The rural development programme in Romania with particular reference to the designation of new towns', Leicester University Geography Department Occasional Paper 13.

Chapter Eight

PROBLEMS OF POLLUTION AND CONSERVATION

Pollution is becoming an extremely serious issue in Eastern Europe, where it is now conceded that environmental damage has not only been inherited from the past but also arisen from the latest technological progress. (1) The high level of gaseous and solid emissions (especially in the more developed northern countries) arises from the emphasis placed on heavy industry and the need to support this effort with electricity generated thermally from low-grade coal. For example in Czechoslovakia 87 per cent of the electricity produced in 1981 (73.5 million kWh) was derived from low calorific coal with a high sulphur content. Official investigations in 1972 showed that 1.60 million tonnes of solid emissions and 2.00 million of gaseous emissions (1.90 million being sulphur dioxide) escaped into the atmosphere of the Czech Lands. Comparable figures for Slovakia were 0.55 and 0.75 respectively. Pollution is widely dispersed since, contrary to earlier suppositions, it is not restricted to one layer of the atmosphere but may rise through several strata, making an impact possibly hundreds of miles from source. Exhalations from the chemical plant at Bratislava cause damage to forests in the Little Carpathians.

However, the problems would be not nearly so great if adequate investment was put into separation plants. More efforts are now being made to control pollution and greater emphasis is being placed on research. The Budapest Polytechnical University has done outstanding work on the survey and mapping of pollution levels. (2) It is becoming clear that the problem requires international action because a great deal of pollution is imported from abroad, a lot of it brought by prevailing winds from western Europe. (3) The situation in southern Poland is made worse by the import of pollution from Czechoslovakia and the GDR (accounting for about half of Poland's air pollution) while virtually all Hungary's pollution originates outside the country: much of the water is already polluted when it crosses the frontiers.

Attempts in Czechoslovakia to measure pollution damage in monetary terms produced very high figures, breaking down into proportions of 60 per cent for health costs, 20 per cent for crops and soils, and 10 each for fauna/livestock and forests. The calculations do not take account of damage to buildings and the general nuisance arising from polluted air and a deteriorating environment. It is quite clear that the health of the population is badly affected by pollution and workers employed in heavy industry are particularly vulnerable. There are numerous cases in Poland, for example, but the Glogow/Legnica copper smelters along with the Nova Huta metallurgical works and the Plock oil refinery are particularly hazardous. And air pollution is additionally injurious to health through reduction in the intensity of the sun's rays, perhaps by as much as 40 per cent. The hazards are manifest in the relatively high incidence of heart disease, lung cancer, and respiratory ailments. The incidence of malignant tumours in North Bohemia, including open cancer of the skin and eyes, is disturbingly high. Infant mortality tends to be higher in polluted areas.

There is no doubt that governments are concerned about pollution: legislation dates back to the 1960s. In 1966-7 general health legislation in Czechoslovakia required all enterprises and organizations to create good conditions of work, provided for economic sanctions against persons and organizations responsible for air pollution, and envisaged an anti-pollution programme supervised by the state. In 1970-1 environment councils were established as part of the government machinery in the Czech Lands and Slovakia. Further legislation has dealt explicitly with water (1973) and forests (1977). Meanwhile in Bulgaria there was an edict in 1960 against air, soil, and water pollution followed by a law in 1963 prohibiting the building and operation of enterprises liable to emit dust and gas: all this was subsumed by further legislation passed between 1967 and 1969. The international dimension is brought out by a conference in Prague in 1971 which brought agreement between the USSR and her six Comecon partners in Eastern Europe (also Yugoslavia in 1973) on scientific and technical co-operation for environmental protection, covering such matters as the rational use of natural resources and the perfection of waste-free technologies. Co-ordinating centres were set up in Bratislava, Dresden, and other cities. Progress was reported to the Stockholm United Nations conference on the environment in 1972 and co-operation has become particularly close between Czechoslovakia and the GDR as regards the water economy.

Medium-term plans now make references to pollution (beginning in Czechoslovakia in 1971-5) but unfortunately the state of the environment does not constitute a plan target. Governments are inevitably ambivalent, wishing to control cases of excessive pollution yet lacking the will to allocate resources on a scale that would constrain the economic development process. Although severe penalties can be imposed on offending enterprises they usually stop short of orders to curtail production and a cynical management will be happier paying fines than building anti-pollution apparatus.

The conjunction of legislative and economic powers in the communist system has proven to be detrimental to ecological preservation in Czechoslovakia. The politico-economic considerations of the communist state have so far always overriden environmental desiderata except in a few minor instances. (4)

At the federal level in Yugoslavia there is a council for the protection and improvement of the environment but the decentralized system complicates the enforcement of federal guidelines which can easily be overlooked by local authorities if the result would be increased costs for self-managed enterprises. (5)

ATMOSPHERIC POLLUTION

Pollution levels tend to be high in the vicinity of power stations burning low-grade coal and also major industrial complexes. (6) Outstanding is the industrial axis running from southern parts of the GDR to Upper Silesia and the Krakow area of Poland as well as Ostrava in Czechoslovakia, and Prague and northern Bohemia, also in Czechoslovakia. In 1977 32.3 per cent of all dust and gas emission in Poland came from Katowice voivodship and a further 33.1 per cent came from a group of five regions (Jelenia Gora, Konin, Krakow, Legnica, and Opole). Deposits of solid matter alone frequently exceeded 1,000 tonnes per square kilometre and occasionally over 2,000 in the early 1970s. The significance of this is all the greater when it is considered that many of the factories responsible for the pollution are situated within or very close to housing estates. In Poland the six worst dust polluters in 1980 were Bogatynia power station (145 thousand tonnes per annum), Oswiecim chemical works (97), Nova Huta metallurgical works (79.1), Jaworzno power station (74.4), the Katowice metallurgical works at Dabrowa Gornicza (62.3), and Nowiny cement works (55.9). For gas emissions the most serious cases again include the installations at Bogatynia (213 thousand tonnes per annum), Dabrowa Gornicza (413.9), Jaworzno (164.9), and Nova Huta (739.0) along with the metallurgical plant Zukowice (360.5) and the Rybnik power station (148.5). The total volume of pollutants discharged into the atmosphere in Czechoslovakia (gases and solids together) rose from 5.2 million tonnes in 1960 to 9.9 million in 1980. But the fall-out is heavily concentrated in certain areas and figures in excess of 400 tons per square kilometre have been reported from Bratislava, Kosice, Prague, and Usti nad Labem. Values exceeding 1,000 tonnes were recorded in Ostrava in 1962. (7)

Outside the main industrial areas of Eastern Europe pollution levels can be very high locally. In the Hron Valley of Slovakia the combination of thermal electricity generation using brown coal, with a range of industries including aluminium, cement, engineering, metallurgy, and wood-processing located in a group of towns including Banska Bystrica, Podbrezova, Ziar nad Hronom, and Zvolen, results in a seriously polluted environment. Air pollution in Hungary affects the hill zone from Miskolc to Veszprem and Ajka, with particularly serious problems in Budapest and the areas around Kazincbarcika, Leninvaros, Miskolc, Ozd, Szekesfehervar and Tatabanya. (8) The pollution involves dust and gas from metallurgical works at Salgotarjan, fine dust from the Labatlan cement works, red bauxite dust at Almasfuzito, and ash and sulphurous compounds emitted at Dorog. In the four Balkan countries the general pollution levels are lower but there are serious problems locally. Sofia lies close to thermal power stations and metallurgical works (Kremikovtsi and Pernik) and pollution has become very severe in other places with heavy industries: Burgas,

275

Table 8.1 Air pollution in Poland 1980

	A	Criteria B	C	D
Poland	7,472.5	100.0	23.9	928
Urban agglomerations	5,480.0	73.3	73.3	585
Agglomerations with the highest emissions	5,017.4	67.2	200.8	368
Of which				
Upper Silesia	2,420.0	32.4	476.0	192
Krakow	984.7	13.2	548.0	24
Legnica/Glogow	529.3	7.1	213.4	8
Rybnik	276.6	3.7	161.4	30
Szczecin	236.4	3.2	60.3	33
Warsaw	198.0	2.6	32.6	36
Opole	186.8	2.5	58.2	16
Bielsko-Biala	185.6	2.5	56.8	29

Source: F.W. Carter (1986) 'Post-war pollution problems in Poland', a paper presented at the Institute of British Geographers, British Polish Seminar, London 1986.

Notes:
A   Total emissions (thousand tonnes/year)
B   Percentage
C   Fall out (tonnes/km^2)
D   Heavy industrial plants

Devnya, Dimitrovgrad, Kardzhali, and Pirdop. (9) In Romania certain cement and chemical works (notably Bicaz in Moldavia and Copsa Mica in Transylvania) create serious local problems, while in Yugoslavia pollution is serious in industrial towns like Bor, Trepca, and Zenica. (10)

Factories and power stations are not the only culprits. Heating units and motor vehicles make a significant contribution to pollution through emissions of various gases including carbon monoxide and nitrogen oxides. Although there are fewer vehicles than in western Europe there has been only limited interest in fitting anti-pollution devices and many of the vehicles are old and poorly maintained. This is all quite separate from the problem of inadequate parking and garaging facilities which results in the cluttering up of streets and recreational open spaces. The railways contribute significantly to pollution where steam locomotives continue in use, perhaps for shunting and freight duties in some of the conurbations. The dust poses a serious problem for street cleaning which is frequently unsatisfactory. The pollution problem is particularly serious under certain atmospheric conditions, particularly when temperature inversions trap polluted air close to the ground. Such a situation occurs in North Bohemia for about 10 per cent of the year and mostly during the winter. Despite tall chimneys (some 300m) the emissions are slow to disperse under such circumstances and are concentrated in a layer of air almost entirely below the level of the mountain tops. This situation is not unique for smog is encountered on calm winter days in many other areas. Outstanding here is the case of Budapest in January 1970 when extremely high levels of nitrogen dioxide pollution gave rise to a medical emergency. Evidently studies of local climatic conditions can be extremely relevant to assessment of pollution hazards through understanding of the convection process, leeward effects, and the circulation of anabatic and catabatic winds.

The response to pollution

In Poland pressures during the Solidarity era were strong enough to bring about the closure of factories. The Skawina aluminium smelter was the outstanding example, built in 1952-4 in the Vistula valley 14 kilometres south-west of Krakow. Production increased from the initial level of 15,000 tonnes to reach 53,000 in 1980: old technological processes were simply concentrated and intensified and with a complete lack of anti-pollution devices more than 1,000 tonnes of fluorine compounds were released into the atmosphere every year. Further pollution came from the power station owing to the ineffectiveness of the filters. Not only did buildings in Krakow suffer from the corrosive effects of hydrogen fluoride but also water supplies were polluted, agricultural output was depressed, and there were dangers to human health. After various attempts to control pollution had proved ineffective public pressure led to the closure of the plant in 1980. Mention should also be made of the cellulose factory at the Chemitex-Celwiskoza chemical fibres plant in Jelenia Gora four years after such action was first seriously mooted (the Bobr river became so polluted as to defy classification), while another factory closure of 1980 was the Analina plant at Lodz, producing viscose fibres, and emitting large concentrations of carbon dioxide and hydrogen sulphide into the atmosphere in the process. There has also been pressure to reduce steel output at Nova Huta.

It is not only in Poland that public expectations have gone beyond the demand for industry at any price. In both Czechoslovakia and the GDR there is considerable public disquiet - so much so that there is an emerging 'green-peace' tendency. Although the political potential may be limited the public can nevertheless exert some influence on the authorities in a way that is difficult to impeach or impede. In the GDR the Evangelical Church has taken tree-planting initiatives and other measures which have attracted enough interest in society to stimulate the party to set up its own society for nature and the environment. In Czechoslovakia Charter 77 has been active on the environmental front, drawing attention to the ecological problems in North Bohemia and social deterioration that may be associated with them, including corruption, alcoholism, and drug addiction. In Central Bohemia one notorious source of ash, the Kraluv Dvur cement works, has been attended to. However this still leaves problems associated with heating plants and small industries in the town. Public disquiet is also emerging in the Balkans. The negative consequences of heavy dependence on lignite are to be seen all over Eastern Europe but perhaps most of all in Sofia where the effect of large power stations and industrial plants in the vicinity was emphasized by the position of the city in a deep and poorly ventilated basin. There has been a big effort in Bulgaria since the 1976-80 Five-Year Plan with increased targets for land reclamation (from mining), tree planting, and establishment of national parks and protected areas. Factory modernization has reduced the amount of pollution: for example the rebuilding of the Damianov copper complex at Pirdop with a recycling of gases previously discharged into the atmosphere. Work has been done to purify the Vit and Yantra rivers, and also the Iskar and Maritsa rivers around the cities of Sofia and Plovdiv respectively. Damage in Bucharest, where dust fall-out may exceed 400 tonnes per square kilometre around industrial estates, has been reduced by the relocation of one notorious offender (the Titan cement works) and the planning of new estates in the southern and south-western parts of the city in which general direction the prevailing wind is blowing. (11) Offensive odours from the slaughter house, carried over long distances under certain atmospheric conditions, have also prompted relocation.

For most enterprises drastic measures like closure or relocation are not necessary since the installation of pollution control devices could be quite feasible. Separation of solid materials (cinders and dust) can be accomplished quite easily but where equipment is fitted it is frequently not properly maintained or modernized. It is reported that a factory exists at Milevsko in Bohemia producing solid separators and filters using a British patent licence. Their equipment has led to reduced dust emissions. Czech and Soviet research has now produced a new filter for power stations; a biological leaching system to remove sulphur from coal is also being developed. But production cannot be increased to satisfy the country's needs because of priority given to other lines of production required for export. For gaseous emissions the equipment is more sophisticated but it seems that the technology is now available. Equipment tested at Tusimice power station near Usti nad Labem traps ash and all but 10 per cent of the sulphur dioxide emissions; and the chemical combine of Lovosice has developed a system which drastically reduces the emission of nitrogen oxides. As long ago as 1970 a scientist at the Czechoslovak Academy of Sciences' Institute of Elementary Chemical Technology designed the world's first third-generation boiler for the desulphurization of low-quality fuels. Patents for this machinery have been applied in the west but the technology has not apparently been used in Czechoslovakia itself.

Pollution from motor vehicles can be reduced by better maintenance and by changes in oil refining, but at a cost to the vehicle owner which is hardly welcome.

The ordinary man is willing to see money spent on filtration plants for industry and may accept that he should pay more for smokeless or liquid fuel in his home. But the automobile is often part of his ego and to its practical benefits is added a deep psychological satisfaction. Even totalitarian regimes must hesitate to tamper with such a situation - least of all in countries where the would-be motorist has only just come to the end of decades of deprivation. (12)

However, pollution in the centre of Budapest is now so bad that walking can be as bad for the lungs as a packet of cigarettes. Traffic will therefore be banned from many streets.

WATER POLLUTION

Pollution of waterways is also serious, especially in the northern countries where it is becoming increasingly difficult to find acceptable supplies of industrial water in the lower sections of river valleys. (13) Chemical and wood-processing industries create the most serious problems in Bohemia as rivers like the Berounka, Bilina, Labe, Ohre, Sazava, and Vltava are now heavily polluted. Problems also arise from mining and smelting operations, from the discharge of sewage, and from fertilizer applications on the land. Pollution of the rivers is also serious in Poland. Many of the major industrial installations are located on the Oder and Vistula while others are situated on the Bug, Notec, Pilica, San, and Warta. There was a marked deterioration during the 1970s and at the end of that decade the worst cases involved two tributaries of the Warta (Ner and Welna) and three of the Vistula (Bug, Bzura, and Utrata). But even on the Vistula pollution has inhibited the increase in fish stocks which were expected to follow the construction of the Wloclawek and other dams. The problem is exacerbated by poor sewage disposal because about half the waste water requiring treatment is discharged straight into rivers, lakes, and seas and barely a fifth receives proper treatment (fewer than half the cities have sewage works).

Research in Hungary, where the Kiskore recreational complex is threatened, demonstrates that even individual industries can create havoc. Chemicals discharged by the food-canning factory at Hatvan have destroyed all life in the Zagyva river, the connection being made tragically clear by the discovery of dead fish covered in tomato juice. The Sajo river is in a similar state: the water is already polluted at the Czechoslovak-Hungarian frontier but further waste materials are discharged in the Miskolc industrial area. Toxic waste has caused many problems in the Hungarian town of Vac: a drug company carelessly disposed of waste and seriously polluted the water supply and on another occasion mishandled combustible materials that could have blown up part of the town. It seems that, albeit ironically, drug companies are notorious offenders. In 1981 a section of a road collapsed in Budapest because a drug manufacturer in Kobanya released so much concentrated waste into the sewers that some of the pipes themselves were damaged.

The problem of water pollution is becoming all the greater because of increased demand for pure water. Water transfers to cities are now required over considerable distances, with plans to take water from the Biala tributary of the Vistula to Katowice and Krakow and even further, over 200km from the Masurian Lakes to Warsaw. At the same time the reduction in forest cover means rapid run-off and consequently a greater need for artificial water storages to assure water supplies. Water purification equipment is available but must be installed more widely with more effective checks against breakdown. Water in the Zelivka reservoir (which feeds Prague) has a nitrate content so high that only bottled mineral water is recommended for mixing baby food. But while the quality of water in the mains occasionally gives grounds for concern, water from wells is invariably polluted. Analysis in Czechoslovakia in 1981 revealed an incidence of pollution exceeding 70 per cent. Use of polluted water has caused epidemics of dysentery, jaundice, and virus-based hepatitis. Similar problems have been reported from Hungary, arising from pollution of ground water through inadequate treatment of sewage.

The water table has been seriously polluted in the vicinity of Prague airport where fuel seepages, arising from poor underground storage, culminated in a major contamination which required nearby villages to rely entirely on mobile water cisterns for a decade. Similar accidents continue to occur as at Slovnaft in Bratislava where, in addition to officially sanctioned daily discharges of 35 tonnes of oily substances into the Danube, there was an accidental leakage of 100,000 tons of oil derivatives in 1974 which affected an underground area of 20 square kilometres. Pollution at Bratislava has threatened Zitny Ostrov, a unique source of ground water capable of supplying the city's needs into the next century at very low cost. But only in 1977 after years of systematic contamination did the Slovak Ministry of Industry endorse the building of a waste water purifying station for Slovnaft and only in the 1981-85 Five-Year Plan was provision made to complete important ecological projects for the rivers in the area. It is regrettable that although technically superior equipment for dealing with polluted water has also been developed in Czechoslovakia it has not yet been widely used. Water treatment can actually be beneficial for the enterprise as in the case of the Sala nitrogen fertilizer factory where ion exchange demineralization stations yield a chemically pure water which can be used by the factory and also a quantity of ammonium nitrate which can be used in the production of liquid fertilizer containing ammonium nitrate and urea. So the importance of safeguarding water supplies is now being appreciated and Czechoslovakia has developed hydro-geological techniques (in collaboration with France) to monitor pollution levels in the Bohemian Chalk Basin which contains a virtually inexhaustible supply of water.

Lake Balaton

Hungary's Lake Balaton is a particularly sensitive area. Before the railway age it was an agricultural region with cereal cultivation on the Mezofold, cattle breeding on the lakeside marshes of Somogy, and viticulture on the volcanic soils of the Bakony Mountains. The development of tourism between the railway line and lake shore was well established by 1914 but intensified between the wars and again after 1945. Established spas like Balatonfured have an excellent local climate and natural aerated springs which are valuable for curing heart diseases. The sanatorium for heart diseases has developed into one of the leading cardiological institutes in Europe. Former volcanic activity is reflected in the geyser craters and distinctive flora of the Tihany peninsula, as well as the fiery Szurkebarat and Keknyelu wines of Badacsony hill. But planning controls have been inadequate:

> the shoreline is now almost continuously built up whereas it should have been used for the construction of beaches, promenades, and resort houses serving the collective types of recreational activities. Moreover architectural style and building regulations have not been adapted to the lake-shore's characteristics. Peripheral areas have also been parcelled out, which has resulted in the construction of thousands of buildings unsuited to the region in places with deficient infrastructural provision. (14)

The whole environment is being undermined by excessive

unauthorized villa development and the problem of heavy tourist pressure is exacerbated by the short season of only three months.

By no means the least of the problems is the progressive eutrophication of the lake, noticed since the early 1970s. The lake is seriously polluted through discharging sewage. Only 15 per cent of homes are connected with the sewage disposal network and a considerable part of the phosphorous originating from individual septic tanks infiltrates into the lake. Moreover river water now includes substantial amounts of chemical fertilizer leached from the soil. Other streams discharge water containing agricultural chemicals (with a significant contribution from water running off the steep slopes on which vineyards have been established) and liquid waste from intensive stock-rearing units. The lake is still satisfactory for swimming but even if further deterioration is prevented it will still take about twenty years to recreate the highly favourable conditions of the early post-war years. Fishing has also been affected by the intensive development of tourism. Fish are harmed by insecticides used to exterminate mosquitoes, and the laying down of concrete to protect the shoreline has deprived fish of their natural breeding and feeding grounds, so that they are less numerous and do not contribute greatly to the biological purification process. Perch and pike have almost died out and must be raised artificially. Meanwhile less palatable species such as bream, gudgeon, and silver carp have become relatively more prominent. The situation has also been complicated by the destruction of reed banks to make room for beaches. With the decline of interest in thatching there has been little economic use for the reeds, yet they contribute to the cleaning of water and must be preserved. The Zala river which used to deposit its alluvium in the marshes now discharges directly into the lake.

Quite apart from ethical questions there are good economic reasons for protecting the lake. Hungary's watering places could not provide a substitute for Balaton if the area became unsuitable for tourism and so both domestic and foreign tourists would be diverted to resorts outside the country with a consequent loss of income and employment. The system of sewage collection and treatment must be improved while better controls must be maintained in agriculture including restriction of further expansion of the vineyards. But it is also important to

re-establish the original filtering function of the reed banks
so that they can absorb much of the phosphorus transported
by the Zala river.

Adriatic and Baltic coasts

Much concern has been expressed about the condition of the
Adriatic coast of Yugoslavia. Sea water is seriously polluted
as a result of the discharge of untreated sewage and also
waste from the principal harbours, including the illegal
dumping of oil dregs. Meanwhile the Baltic is becoming one
of the world's most polluted seas and Poland contributes
disproportionately to this process with a sixth of the total
pollution, much of it in the form of sewage from more than
eighty outlets. (15) Serious pollution now occurs in the
Gdansk area where water is affected by sewage (treatment
plant is either inadequate or non-existent), waste from local
industry and shipping and impure Vistula river water
containing chemicals discharged at the Plock refinery. The
coastal environment is also suffering from excessive litter
and erosion of the fragile vegetation on sandy soils through
heavy tourist pressure. A landscape park was set up in 1979
for the coastal strip from Gdynia to Wladyslawowo and the
Hel peninsula but the conflicts continue because of serious
pollution arising from the lack of sewage works and the
complication of plans to start fish-farming. The position
became serious enough for the authorities to close most of
the beaches in Gdansk Bay and Puck Bay. But now the water
management companies are building regional sewage
treatment plants in Elblag, Gdansk, Gdynia, Stegny, and
Swarzewo. Improvements are also being made along the
Vistula valley but it will be possible to reopen the beaches
only in the 1990s. However, some improvements are already
evident: the sea water is less polluted, pollution-sensitive
species are increasing, and fish meat contains fewer harmful
compounds. An organization has been set up to deal with
emergencies and this was very effective in 1986 in dealing
with oil spills in the bay of Gdansk and in the Odra river. In
the near future sewage treatment should be improved in all
major cities in Poland and waste water treatment should be
carried out at the principal chemical and metallurgical
plants.

POLLUTION AND CONSERVATION IN THE COUNTRYSIDE

Air pollution can damage crops and cause a decline in soil fertility through salinity. But the use of fertilizers, herbicides, and pesticides can easily become counter-productive. With the continual degeneration of the soil damaged agricultural land may be immune to further chemical boosting. As much as a quarter of the fertilizer may then pass through the soil to reach the streams where it acts as a pollutant. Herbicides and pesticides represent a danger: the residual content of these substances in farm produce and feeds is not being tested systematically and the long-term ecological consequences may be serious. Other forms of pollution include biological gases escaping from large-scale pig-breeding units, careless air spraying, inadequate retention of silage juices, the release of lubricants and diesel oil from agricultural machines into rivers and streams, and the washing of tractors with detergents in natural water. These problems are already serious in some areas and are likely to become more so in future. Some useful work is being done at the experimental farm of Dolna Malanta near Nitra in Slovakia where agriculture and forest management is based on reduced consumption of chemical fertilizers and pesticides.

Most dramatic, however, is the loss of agricultural land through the dumping of waste. (16) As open-cast working of low-grade coal involves greater output and greater depth of working the amounts of over-burden are unprecedented. With the ratio of over-burden to coal in Czechoslovakia being now 5:1 the area covered by waste has risen from 800 ha in 1970 to 1,800 in 1980. Reclamation is being given more attention and it is evident that in North Bohemia, while the amount of waste land increased from 145 square kilometres in 1970 to 205 in 1980, the area reclaimed increased from 61 to 128 square kilometres. Information from Poland indicates an increase in the amount of devastated land from 74.0 thousand ha between 1973 and 1979 to 107.4 between 1980 and 1984. The annual reclamation effort remained unchanged at 4.5 thousand ha per annum. However only modest areas can be returned to agriculture (20 square kilometres in Czechoslovakia by 1980) because the mixed soil on the waste heaps is quite unsuitable. It seems that the best that can be expected is that half the reclaimed land will be used for agriculture and the other half for forest parks, recreation facilities, and building. For example the

285

ceramics works at Lubna near Rakovnits in Central Bohemia, which started producing tiles in 1982, is sited on a former shale mine. In the GDR the closure of lignite quarries in the Geiseltal has provided some scope for water-based recreation (and the building of weekend cottages) and now that the copper smelter at Eisleben is processing ore from Sangerhausen alone (since 1972) the waste heaps that have for long been a symbol of the Mansfeld copper industry are being cleared away. But there is careful documentation to maintain an accurate record and at least one tip will remain as a permanent reminder of the former importance of the industry.

Damage to woodlands

The situation in the woodlands of Eastern Europe has become very serious. Sulphur dioxide induces acid rain which caused some tree deaths in North Bohemia as early as 1947. But by the 1960s serious damage to forests was reported in the whole border region between Czechoslovakia, the GDR and Poland, with 100,000 ha of dead woodland in the Krusne Hory alone. During the 1970s sulphur dioxide emissions, coming mostly from lignite-burning power stations, reached 37 tonnes per square kilometre (compared with 14 in the FRG where more stringent controls are in force). Damage is particularly severe in the Ore Mountains (Erzgebirge), sandwiched between power stations and chemical plants using brown coal in Czechoslovakia and the GDR. Great damage has been done in the Jizerske Hory Mountains with particularly heavy precipitation. The Giant Mountains (Krkonose/Riesengebirge) are also highly vulnerable on account of power stations in Czechoslovakia (Chvaletice and Opatovice near Hradec Kralove), brown coal processing in the GDR (Schwarze Pumpe near Hoyerswerda), and metallurgical works in Poland (Upper Silesia and Krakow). The situation in the Orlicke Hory Mountains is particularly serious. There is also concern about the forests on the border of Moravia and Slovakia and those along Czechoslovakia's frontier with the FRG and it is feared that within twenty to thirty years the mountains could be completely bare.

Damage is also expressed through the depradations of pests which increase in numbers according to pollution intensity. One of the worst pests is the nun tussock moth which was established over a forest area of 40,000 ha in 1978 rising to 0.5 million ha in 1980 and approaching 2.0 million in 1981. Trees weakened by pollution and disease fall easy victim to gales and snowstorms so they have to be cleared away or treated with chemicals, liberally air-sprayed over the Jizerske Hory, Krkonose, and Krusne Hory forests in 1979 and 1980. In the Karkonosze Park a freak storm in 1966 destroyed 300 ha of forest which succumbed to the dreaded bark beetle: by 1977 this pest had attacked an area three times larger than the one originally affected. Damage is most marked in coniferous woodlands and steps are being taken to replant the worst affected areas using more resistant species with good growth qualities (some of them imported from North America). However the exotic trees have not yet been used successfully and in any case the best that can be expected is a reduction in the rate of degeneration rather than a halt to the process altogether. Meanwhile deciduous woodlands are being affected and animal life is disappearing as the woodland thins out.

Protection measures are gaining momentum and in Czechoslovakia it is now common practice to spray woodland every two years using ground limestone (containing some magnesium) to reduce the soil acidity caused by sulphur dioxide emissions. However, the forests must be protected not only from pollution but also from heavy felling which is carried out to maintain production in the wood-processing industry. Many forests are worthy of preservation because of their particular ecological interest. For example, the Bialowieza forest on the Polish-Soviet border near Bialystok where one of Europe's most important ancient lowland forests extends over some 1,250 square kilometres (580 in Poland) including much marsh and peat bog. A rich variety of species is included in the canopy climax vegetation. However the area of woodland should really be increased both in the countryside and in the cities. Green belts are a prominent feature of the latest urban plans. They have an obvious visual impact while providing recreational open space as well as 'lungs' which make for less polluted air, lower temperatures, and higher humidity levels. Such belts will protect the new housing complexes planned for Prague and will also prevent the city from extending all the way to neighbouring towns like Beroun, Brandys, Kladno, and Kralupy. Similarly in Bratislava green belts will also contain the three new residential zones and insulate the city's major industrial plant, the Slovnaft refinery. But at the present time the parkland in Bratislava is much too small to meet the needs of the population and to exert an optimal effect on the micro-climate. Woodland also plays an important role in the Ajazmoto park in Stara Zagora, one of a growing number of urban parks on the edge of Bulgarian cities.

The significance of woodland is also appreciated in Budapest. (17) A comprehensive environmental act was passed in 1976 while a National Environmental Protection Concept was drawn up in 1980 and applied to Budapest a year later. The development of the city in the period up to 2000 will involve careful siting of green spaces in order to dissipate the intensity of the urban heat island. Green spaces of up to 100 ha, comprising overall some 20-30 per cent of the total urban area can have a significant moderating influence. Well-forested green spaces can be particularly effective. Budapest as a whole is well off since biologically active surfaces make up about half the total area of the city but the distribution is very uneven with very few green spaces in the core. Moreover too much parkland is covered with a gravel surface and trees on road-side verges are not properly looked after: they are often destroyed in the first stages of redevelopment even though replacements need twenty years before a perceptible micro-climatic influence is exerted. The Buda Hills are very valuable because they offer cool conditions in summer, long hours of sunshine in winter, as well as clean air throughout the year. The dominant north-westerly winds and the night breeze produced by the circulation of air in the hills and valleys mean that the Buda Hills have a positive effect on the climate of the inner city. However, large-scale post-war housing developments have impeded air circulation and destroyed a great deal of hill-slope vegetation. There is now a better appreciation of the ecological issues although the planners have gone no further than to restrict the size of plots for further housing development and control the quality of domestic fuel in order to minimize sulphur emissions in the Buda Hills themselves.

Protection: nature reserves and national parks

While control of pollution at source is obviously important the other side of the coin involves the protection of unique environments. These may include specific areas of woodland

and farmland or particular buildings and complexes. The formation of nature reserves and national parks dates back to the inter-war years. In Bulgaria a council for the protection of the countryside was founded in 1928 thanks to the encouragement of the Bulgarian botanist Stefan Petkov. Reserves were subsequently created in the Silkosija area of the Strandzha Mountains (1931) and the Parangalitsa area of the Rila Mountains (1933). A national park was formed in the forested Vitosha Mountains near Sofia in 1934. This now extends over 25.6 thousand hectares and is the most popular recreational area in Bulgaria. By 1940 4,200 hectares of land were protected, but momentum was lost by the war and by subsequent industrialization which had a negative effect on the Vitosha park: although providing 'lungs' for Sofia the environment was badly affected by pollution from power stations (such as Pernik) and metallurgical plants like Kremitovtsi. In 1960 the first state decree for the protection of the countryside scheduled an area of 18,000 ha since when great progress has been made.

Many nature reserves have been set up since 1969 when the main priorities were identified. All the six large reserves (exceeding a thousand hectares) and most of the small ones (100-1,000 ha) are in the mountains where various rock formations, plants, and animals have been protected. But the smaller reserves in the lowlands are also notable: Arkutino on the Black Sea coast comprises mud flats overgrown with water lilies and papyrus rushes. There are now ten national parks although almost 90 per cent of the total area involved falls to the two largest parks of Pirin and Vitosha, the latter being extremely important for the people of Sofia who flock to the park at weekends. However the other reserves have a value out of all proportion to their area, notably the Ropotamo and Zlatni Piassatsi parks on the Black Sea coast which are interesting in terms of landforms and vegetation. The former, situated near Burgas, includes the Arkutino reserve and three others, all of them offering important botanical resources. In 1985 a total protected area of 176.5 thousand hectares (0.2 per cent of the country) was made up of ten national parks (70.8 thousand hectares), 98 nature reserves (54.7), and 64 protected sites (22.0) as well as 942 historical places (11.9) and 2,197 sites of outstanding beauty (22.0). Attention is also given to the protection of certain plants (sea and water lilies for example) in all parts of the country and various birds and animals (the loss of bustards in Dobrudzha through

Figure **8.1** Nature reserves in Romania

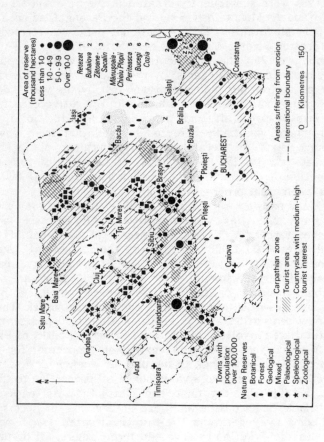

Source: Atlas RSR (Bucharest 1979).

Figure 8.2 Nature reserves in the Danube Delta, Romania

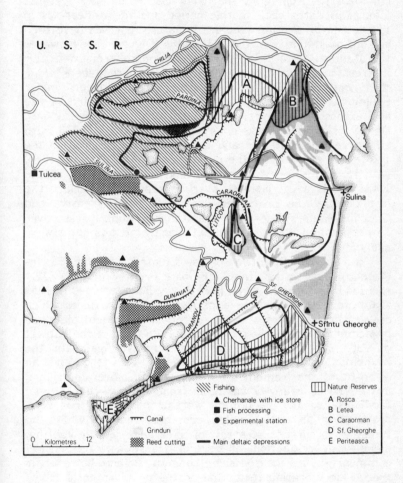

Source: D. Turnock (1979) 'Water resource management problems in Romania', <u>GeoJournal</u> 3: 609-22.

291

intensive farming and tortoises through smuggling being matters of particular concern). The importance of the swamps of Burgas for bird migrations is also appreciated.

Elsewhere important initiatives were taken in the inter-war years with strong pressure exerted by the scientific community. Romania's national park in the Retezat Mountains dates back to this period and the network is in the course of development. Many nature reserves have been set up, particularly in the mountains (Figure 8.1). However the Danube delta is a lowland environment of particular interest and while the exploitation of the reed beds along the northern (Chilia) distributary has induced major ecological changes (affecting the habitat for the birds) there are extensive reserves in the southern part (Figure 8.2). Hungary's national parks embrace the Hortobagy (1973) with an interesting wild life and an ethnography related to the traditional economy of the puszta, and the Kiskunsag (1975) with its relict forest (Toserdo), and a dead branch of the Tisza at Tiszazug, isolated by river regulation work in the last century. So protected areas are now extensive and Czechoslovakia's 800 reserves and four national parks cover 15 per cent of the country. But management is not always adequate.

In Poland the two parks situated on the Polish-Czechoslovak border, the Tatra park created in 1954 (22.1 thousand hectares) and the Karkonosze Park created in 1959 (5.6), are both subject to excessive tourist pressures which are destined to increase when current development plans are implemented. It will be more difficult to protect the habitat of rare species like eagles and falcons. Again, in the Kampinos Park of 1959 (involving a 22.3 thousand ha remnant of the Kampinos Forest extending over marshland in the Vistula valley downstream from Warsaw) recreational facilities for the people of Warsaw have still not been provided in an ecologically acceptable manner while the forestry authority's strategy of clearance and replanting, along with relocation of several traditional agricultural communities, has caused some concern. Poland, however, is greatly accelerating its action on the environmental front. There are over 900 nature reserves and more than 10 per cent of them were designated between 1983 and 1985 alone. An even more rapid improvement was made during this period in the creation of new protected landscape areas, doubling the total area from 1 to 2 million hectares. When taken along with resolute action over the treatment of sewage and control of gas emissions by filtration this represents a significant break-through.

Protection: scheduled buildings

It is also recognized that rural architecture represents certain values and should be preserved. Hence the importance of legislation to provide protection for listed monuments. Action generally dates back to the 1950s and 1960s but it has usually been necessary to follow up with more effective measures like the 1973 Czechoslovak law for the protection of historical objects, which superseded the first law on the protection of historic monuments, passed in 1958. Great importance has been attached to castles and chateaux which were taken over by the government in 1945. In 1948 about 150 were selected for opening to the public with interest attaching to the contents of the buildings and their parkland setting as well as the structures themselves. A landscape park in the French Empire style is being maintained at Chroustovice while nineteenth-century English parks are retained at the castles of Horsovsky Tyn and Slatinany near Chrudim as well as the hunting lodge of Kozel. The latter was built in the late eighteenth century complete with riding hall, stable, and chapel and provides an excellent illustration of life at a country seat in the last century. It is also noteworthy that the Botanical Institute of the Czechoslovak Academy of Sciences at Pruhonice is located in the gardens improved by Count Silva-Tarouca after the estate was acquired in 1885. The land was taken over by a state experimental institute in 1927 and the improvement of the park continues. Czechoslovakia also provides examples of monuments which have been dismantled and rebuilt on new sites when threatened by hydro-electric schemes. The suspension bridge built across the Vltava at Podoli near Prague in 1847 was rebuilt at Luznice near Tabor when it became threatened by the Orlik dam in the 1960s. In other parts of Eastern Europe churches and monasteries have been threatened. The hydro-electric project for the Trebisnjica valley in Yugoslavia threatened the church belonging to the Piva monastery and so it was rebuilt on a higher site safe from the rising water level.

Museums have an important place in conservation work for the rural areas. A network of ethnographical museums is developing in Czechoslovakia. Locations include Knourin,

Martin, and Roznov pod Radhostem, the latter dating back to the 1920s. It is usual to include open-air sections with a range of churches, cottages, mills, and workshops. At Kysuca near Nova Bystrica the complex includes a working narrow-gauge railway formerly used for the transport of timber. Other museums have more specific functions like the brewing museum opened in Plzen in 1959 and the museum of the Andrassy family opened at Krasna Horka near Roznava. The other East European countries are making similar efforts to develop museums. Outstanding examples include the village museum in Bucharest, started by sociologists during the inter-war period, and the Etura ethnographic museum south of Gabrovo in Bulgaria. This contains eighteenth- and nineteenth-century houses, brought from the old quarter of Gabrovo and from adjacent villages, and a number of workshops concerned with wood, leather, metal and ceramics. The idea emerged locally out of interest in old industrial premises: a workshop for blacksmiths and cutlers, and a primitive water-powered corn mill using the water power from the Seveka and Yantra rivers.

Monument protection in Hungary was provided for in the law of 1949 but progress has been particularly substantial since the Ministry of Building and Construction set up a Monument Protection Board in 1957 and issued building regulations covering monument protection in 1961. Various palaces belonging to the former Hungarian aristocracy have been restored including those of Magocs (at Pacin), Melczer (at Keked), and Rakoczi (at Szerencs). The Hungarian Renaissance style contributes to their value while current institutional use provides good conditions for adequate future maintenance. Attention is also given to the restoration of castles such as Coburg (Edeleny), Eszterhazy (Oroszlany), Grasskovich (Godollo), Karolyi (Fehervare-surgo), and Zichy (Obuda); also the ramparts, walls, and forts at Buda, Eger, Esztergom, Sarospatak, Veszprem, and Visegrad. Conservation work in Hungary also includes several settlements with valuable folk architecture (Csongrad, Fertorakos, Halloko, Szigliget, and Tihany) including many houses relating to the national minorities. There are also national museums in Bekescsaba (Romanian and Slovak), Mohacs (South Slav), and Tata (German). Relics of agricultural history attract attention: more than 200 buildings have been placed under protection including farmsteads, mills, barns, and wine cellars. Roughly similar action is being taken in the other countries. Care is taken to preserve houses in some old Bulgarian villages while the Orthodox monasteries (with exterior frescoes) have been preserved in the Suceava area of Moldavia. The Romanians are also very keen on ethnographic museums which date back to the inter-war years. Those in Bucharest, Cluj-Napoca, and Sibiu have major outdoor sections where buildings are collected together from across the country.

CONSERVATION OF HISTORIC TOWNS

The need for conservation in the towns of Eastern Europe is overwhelming. There is an extremely rich architectural heritage. (18) In Yugoslavia there is the Islamic architecture of Macedonia well seen in the centres of cities like Bitola, Prilep, and Skopje where mosques, inns, and houses form distinctive townscapes which should be retained despite the ever-present threat of earthquake as well as modern commercial pressures. By contrast in the Dalmatian cities like Split it is the Roman legacy along with a medieval/early modern succession from Romanesque to Gothic, Renaissance, and Baroque which merits attention. However the historic buildings of Eastern Europe suffered considerable war damage and since 1945 the environment has frequently deteriorated because suburban developments have increased pressure on the central areas. (19) For example heavy traffic causes noise, congestion, and damage to buildings on account of vibration and gaseous exhaust emissions. Reconstruction of historic buildings has been going on throughout the post-war period but at an increasing scale as pollution damage has increased. Such work depends on the creation of effective organizations under the control of the appropriate ministry (usually the Ministry of Culture). In Czechoslovakia a state institute for the reconstruction of historic towns was set up in Prague in 1954. Early experience was gained through the restoration of the centre of Cheb. The aim is to restore old buildings and find suitable new functions for them (thus the famous Convent of the Blessed Agnes is now used as a national art gallery) and close liaison is maintained with the International Council of Monuments and Sites, affiliated to UNESCO. Operating in a similar international context Poland has set up a total of eleven state monument preservation workshops based in the main cities.

Progress is, however, constrained by several factors. First there is some ambivalence on the part of the authorities towards historic buildings. Restoration of the Royal Castle in Warsaw was projected as a taste of the first Five-Year Plan but a negative attitude to the relics of feudalism (the oldest parts of the castle are fourteenth century) led to postponement until 1971 when work went ahead with the support of Poles from all over the country. In Bucharest at the time of writing some important churches are being lost through an ambitious redevelopment project which is replacing the buildings of the Uranus district with a new boulevard, and associated administrative, commercial, and residential buildings, commemorating the victory of socialism. Then cost is always a problem and progress of major projects is inevitably slow. In Budapest a lot of attention has been given to the Castle Hill district of Buda including the royal castle, the medieval Buzoganytorony (Mace Tower), Matthias Church, Halaszbastya (Fishermen's Bastion), and various villas and mansions in the Buda Hills. Work has been going on for years and recently the construction of new hotels has acted as a stimulus to renovate old buildings in the immediate environment.

Conservation must be organized in the context of significant levels of private ownership of property in town centres, especially in the northern countries. And the problem may be compounded by the immobility of the owners on account of age and low income. Also by the tight control of rents so that income received from tenants is usually too small to keep the property in a good state of repair. In Poland it is only in the northern and western 'recovered' territories that the state passed special legislation to gain ownership of urban land, so that the majority of buildings in the centres of cities like Krakow and Torun are privately owned. In Czechoslovakia much landed property has been nationalized in all parts of the country but significant contrasts still remain in urban centres: while Ceske Krumlov has virtually no private houses almost a third of Tabor's smaller buildings are privately owned. By contrast in the Balkans nationalization of land ownership has been more comprehensive and in Yugoslavia the constitutions of all the constituent republics state that no one may own land in cities.

Restoration work has been geared to individual buildings. For example in Hungary a major effort was made during the 1970s to restore the nineteenth-century cathedral of Esztergom which includes the remarkable Renaissance Bakocz Chapel of the sixteenth century. And several other historic churches have also been reconstructed including the catholic church in Balatonszolos and the Protestant churches at Adorjas, Miskolc and Nyirmihalydi; also the episcopal palace in Veszprem. However, there is also concern for urban districts where the whole area needs to be taken in hand. Immediately after the war the decision was taken to rebuild the historic centres of Gdansk and Warsaw and in later years the conservation of other valuable townscapes which had survived the war relatively unscathed has been an important priority, as at Krakow and Zamosc: such areas are now given over to housing and also to arts and crafts which have important tourist interest. In Hungary the centres of sixteen towns and five villages have been taken into care. The centre of Sopron was declared a conservation area in 1965 and nearly two hundred buildings have since been restored over an area of 30.3 ha. Careful archaeological work has been carried out in the process and the creation of a pedestrian precinct allows the environment to be appreciated without the distraction of vehicular traffic. The achievement has won international acclaim. Sarajevo has its famous bazaar of Bascarsija enhanced by four notable examples of medieval oriental town architecture: the Bascarsija mosque, the Morica inn (han) and shopping arcades (bezistan) of Brus and Gazi-Husref Bey. Nearby is the Kursumli school (medresa), the oldest educational institution in the region. Restoration in Bulgaria may be seen at Trnovo and Koprovniste.

Restoration work in Czechoslovakia has moved forward from attention to individual buildings on an ad-hoc basis to systematic programmes for entire town centres, following the law of 1958 which created a legal basis for the care of historic towns. There are now thirty-five historic town reservations in Czechoslovakia. Interest focuses primarily on the medieval street system and the traditional buildings. Some 200 houses are involved at Ceske Krumlov, one of several towns in Czechoslovakia which has received international recognition for its historic interest. Renovation on a similar scale has been going on at Cheb and Karlovy Vary: also at Telc with its funnel-shaped town square (reflecting the widening corridor of land leading away from a medieval moated fortress) lined by burghers' houses with continuous arcading. (20) Arcading is also prominent around the square at Ceske Budejovice, a bastide

297

town built at the confluence of the Vltava and Malse to consolidate control of the border region. However in the typical case the conservation programme includes the castle which, as a feudal fortress planted in what would have been originally a rural settlement, provided the initial stimulus for urban growth. Thus at Mikulov Henry of Liechtenstein's Gothic castle of the fourteenth century was rebuilt in the Renaissance style in the seventeenth and again in the Baroque style following an eighteenth-century fire. And in the Hussite town of Tabor, fortified through the fifteenth century, the aim is to clear the hillside of recent buildings so as to reveal the circuit of medieval walls and bastions. Church buildings are prominent in several restoration plans, especially Kutna Hora, with the Cathedral of St Barbara the key building in this medieval mining town.

Krakow (Poland)

Krakow is of the greatest importance for Poland on account of its historical role. The Wawel cathedral remained the royal burial place until the end of the Polish monarchy in the eighteenth century. The Jagiellonian University (founded 1364) has a history of more than 600 years of scholarly work: as the seat of the only Polish language university during the partition period it became the vital centre of cultural life prior to 1918. But there is also a wealth of important buildings. The castle and cathedral on the Wawel Hill overlook the main market-place (Glowny Rynek), with the medieval cloth hall (Sukiennice) and the church of St Mary's, and the small market (Maly Rynek). The medieval core is made up largely of buildings of three or four storeys built along narrow streets which are clearly bounded by a wooded belt (Planty) laid out during the 1820s on the site of medieval fortifications. A late-nineteenth-century defence system then created another distinct morphological zone for advance further into the countryside was constrained until 1906.

Krakow emerged from the Second World War undamaged, apparently because of an ingenious manoeuvre on the part of the Red Army. It was the only major city in post-war Poland with its infrastructure reasonably intact. However, while heavy investment went into Gdansk and Warsaw, Krakow was left unattended and rapid deterioration took place. Particularly significant was the decision of 1949 to site the Lenin steel works on the eastern edge of the city along with a huge suburban housing complex (Nova Huta) to generate a proletarian community, with a population which is now some 250,000. Today there is a population of some 0.12 millions within the 5 square kilometres of the core (density 24,000 per square kilometre) in contrast to the population of 0.49 million distributed at a density of 4,100 per square kilometre over the 121 square kilometres of suburbs. Finally there are 197 square kilometres of land within the administrative limits of Krakow which constitute the fringe zone: 0.07 million people live in a network of villages at an average density of just 370 per square kilometre: modern detached three-storey houses are tending to replace traditional single-storey wooden farmhouses. Overall the population of the city continues to rise and should advance from 0.75 million in 1975 to 0.90 in 1990 and 1.00 in the year 2000. (21)

In the 1960s the local authorities were urged to take the necessary steps to preserve the city but comprehensive plans did not take shape until 1974. Then in 1978 UNESCO announced that Krakow, along with the neighbouring salt mine of Wieliczka, was one of the twelve finest monuments of nature and civilization in the world. The main problem is the air pollution: 0.17 million tonnes of dust and 1.20 million tonnes of gases (sulphur dioxide and carbon monoxide) fall on the city each year. Most of it comes from the Nova Huta iron and steel works, but industrial installations in Upper Silesia, including the Katowice foundry and other units in Chrzanow, Jaworzno, and Oswiecim, also affect the city when winds blow from the west. Then there are thousands of local boiler houses and domestic heating systems, all using low-quality coal. The result is the worst kind of smog which leads to heavy damage to historic buildings. Furthermore, the foundations of houses are being eroded by underground water relating to an intricate system of canals and tunnels whose exact location is a matter of conjecture since the Germans destroyed the plans to frustrate the Polish underground during the war. Entire blocks are threatened with collapse if their foundations are not reinforced. Finally, a grave threat in the future is posed by the proposed Vistula barrage which could lead to a higher water table and the flooding of cellars in the old houses.

The protection of Krakow calls for a concerted effort. Anti-pollution devices should be installed at factories and green belts should be established around the city. In the heart of the city there should be a switch to gas heating and closure of streets to motor traffic combined with redirection on to new orbital highways further from the centre. In the core an important conservation programme began in 1974. The ownership structure has been left intact although compulsory purchase could take place where the cost of renovation exceeds half the property's total value. This contrasts with the attitude taken by the authorities in Torun where all restored buildings have passed into state ownership. Work proceeds block by block with a typical block containing some twenty house sites involving almost as many separate owners. About a third of the houses are beyond repair, necessitating demolition and rebuilding, while other houses require only attention to plasterwork along with exterior painting and inside modernization. Restoration is expensive however and provision of adequate funds is the acid test of political will, enabling skilled workers to be recruited and building materials provided. Obviously state funds are essential but some contributions are made by local government and by enterprises while the public at large may be exhorted to contribute: support for conservation work in Krakow has been spelt out as a patriotic duty for all Poles including those now living abroad. But progress is also retarded by the difficulty of emptying whole blocks at a time, given the predominantly elderly population. Conservation is therefore becoming a race against time as the decay of the stone work continues at a rapid rate. Gloomy pictures of crumbling statues, dislodged balconies, and shored-up streets will persist for some time to come.

Prague (Czechoslovakia)

Prague also poses a challenging conservation problem with its rich medieval heritage. This focuses first on Hradcany, built on the Vysehrad hill overlooking the Vltava. Here the medieval cathedral and castle dominate a site given over largely to the functions of government. On the low ground a trading city (Stare Mesto) developed within a bend on the river with a system of defences and a bridge, the Charles Bridge (Karluv Most), connecting the trading settlement with Mala Strana on the opposite bank of the river. The total wealth of the monuments is incalculable with outstanding examples for a range of architectural styles extending from the Romanesque, like King Sobeslav's twelfth-century palace, Early Gothic such as the Convent of the Blessed Agnes of Bohemia, High Gothic exemplified by Charles Bridge, Late Gothic typified by Tyn Church and the Powder Tower, Renaissance evidenced by the Thun Palace, and Baroque by the Sylva-Taroucca Palace. The sequence continues with the Empire, Romantic Gothic, and Neo-Renaissance periods reflected, for example, in the Kinsky Villa, Baumgarten Summer Palace, and National Theatre respectively. The value of this heritage rests not only in the individual structures which have been given considerable attention since the 1960s (a total of more than 1,700 monuments were catalogued by 1964) but in the whole ensemble: hence the declaration of 750 ha of central Prague as a monument reservation in 1971. Special attention has been given to the old commercial town of Nove Mesto. Progress is being made with restoration yet the pressures remain extremely severe in view of the high-population density, rapid industrial development, and heavy dependence on low-grade coal. High dust fallout has been substantially reduced by greater use of hydrocarbon fuels imported from the Soviet Union and the problems of noise and vibration arising from heavy traffic may be moderated by the expansion of the metro system. This should cater for new suburban developments to the south and south-west of Prague which will have upwards of 100,000 inhabitants each. Meanwhile the most modern technology is being used to monitor the continuing deterioration of key buildings so that the most pressing problems can be attended to. (22)

Dubrovnik (Yugoslavia)

Dubrovnik is another medieval gem, for the economic prosperity of this Adriatic trading settlement went hand in hand with cultural progress seen in literature, music, technology, and inevitably architecture especially during the fifteenth and sixteenth centuries. Once again the interest is international, underlined by UNESCO's recognition of the town (along with Kotor and Split) as one of great importance for conservation. There is a planned town, mainly of late-thirteenth-century date, with a massive system of walls and fortifications which developed through to the seventeenth century. The buildings include palaces such as the Rector's Palace (Knezev Dvor), the cathedral (Katedrala), religious buildings such as the Franciscan church and monastery

(Samostan Male Brace-Fratres Minores), shops including one of Europe's oldest pharmacies (Franjevacka Apoteka), and fountains such as the Jewish fountain (Zidovska Cesma) which supply water brought from springs deep in the karstic hinterland. But problems for Dubrovnik include the liability to earthquakes (which have occurred in this century alone on two occasions) and the corrosive action of sea salt on the limestone buildings leading to severe weathering of the more exposed monuments, especially the city walls.

CONCLUSION

The conservation problem in Eastern Europe is certainly a serious one and it is inevitable that further serious damage will be done to the cultural heritage. The total number of scheduled monuments (only 8,500 in Hungary) is low by European standards and restoration is proceeding only slowly. So there must be continuing decay and deterioration of those monuments selected for preservation (to say nothing of other monuments which will be lost through further development). On the other hand, despite certain aberrations like the negative attitude shown recently in Bucharest, there is a growing commitment by the authorities and the population at large is more highly motivated than ever before. Moreover, although it is not without its own hazards, there is the prospect of nuclear power providing the ultimate solution to the pollution problem, through reduced consumption of low-grade coal. Hence the remark that 'the people's democracies of Eastern Europe seem to have condensed environmental abuse into a shorter timespan than the West and real attempts to reverse the situation were evident everywhere in the 1970s'. (23) So there may still be time.

NOTES

1.   Fullenbach (1981); Singleton (1987). See also L. Dienes (1974) 'Environmental disruption in Eastern Europe', in Volgyes (ed.) 141-58; L. Dienes (1974) 'Environmental disruption and its mechanism in Eastern Europe', Professional Geographer 26: 375-81; I.P. Gherasimov (1974) 'The present state of the natural environment in Europe', in Pecsi and Probald (eds) 89-97; J.M. Kramer (1983) 'The

environmental crisis in Eastern Europe: the price of progress', Slavic Review 42: 204-20; I. Oldberg (1983) 'Planned economy and environmental problems: Eastern Europe from a comparative perspective', Bidrag till Ostatsforskningen 11: 1-64.

2.   G. Enyedi (1974) 'Planning for purposeful use of the environment: a Hungarian viewpoint', in Volgyes (ed.) 123-9; D. Kulcsar (1984) 'Environmental quality control policy in an economic and regional context', in Compton and Pecsi (eds) 155-66.

3.   H. Karrasch (1983) 'Transboundary air pollution in Europe', Heidelberger Geographische Arbeiten 73: 321-44; A. Rosencranz (1980) 'The problem of transboundary pollution', Environment 22(5): 15-20.

4.   F. Pohl (1983) 'Environmental deterioration in Czechoslovakia', Radio Free Europe Background Report 95: 15.

5.   F. Singleton (1985) 'Environmental problems in Eastern Europe', in L. Collins (ed.) Anglo-Bulgarian Symposium 1982, London: University of London School of Slavonic and East European Studies 2: 58-74.

6.   Z. Chojnicki (1972) 'An economic approach to some problems in using the geographical environment', Geographia Polonica 20: 42-7; S. Gilewska (1964) 'Changes in the geographical environment brought about by industrialization and urbanization', Geographia Polonica 3: 201-10; J.W. Zwozdziak and A.B. Zwozdziak (1985) 'Evaluation of atmospheric trace species in the vicinity of a copper smelter and a power plant', International Journal of Environmental Studies 24: 97-105.

7.   F.W. Carter (1985) 'Pollution problems in post-war Czechoslovakia', Transactions Institute of British Geographers 10: 17-44.

8.   S. Katona (1979) 'The assessment of anthropogenic effects on environment in the Budapest agglomeration', Geograficky Casopis 31(1): 12-27; E. Probald (1974) 'Air pollution and the urban climate of Budapest', in Pecsi and Probald (eds) 191-6.

9.   H. Tichkov (1974) 'Thermal inversion and industrial pollution of the near-surface air of some depressions in Bulgaria', in Pecsi and Probald (eds) 197-202.

10.   D.E. Kromm (1973) 'Response to air pollution in Ljubljana', Annals Association of American Geographers 63: 208-17; D.E. Kromm (1974) 'Perception of air pollution hazards in Ljubljana, Yugoslavia', in Volgyes (ed.) 130-40; M.

Pasic (1972-3) 'Problems of environmental protection in Yugoslavia', Slavic Papers 6-7: 90-100.
11. C. Herbst et al. (1974) 'Correlation between the industrial zones of Bucharest city and the pollution of its atmosphere', in Pecsi and Probald (eds) 187-90.
12. Radio Free Europe Research Report: Eastern Europe 7 (1974) :11.
13. J. Szupryezinski (1976) 'The effect of the reservoir near Wlockawek on the geographical environment', Geographia Polonica 33: 135-41; D. Turnock (1979) 'Water resource management problems in Romania', GeoJournal 3: 609-22; D. Turnock (1982) 'Romanian geography reunited: the integrative approach demonstrated by the conservation movement', GeoJournal 6: 419-31; M. Zajbert (1975) 'The problem of use and protection of water resources', Geographia Polonica 32: 63-74.
14. G. Bora (1984) 'Environmental management in the Lake Balaton region', in Compton and Pecsi (eds) 91-108 (quote p. 97).
15. E. Leppakoski (1980) 'Man's impact on the Baltic ecosystem', Ambio 9: 174-81.
16. K. Billwitz (1975) 'Some natural environmental changes within the Halle-Leipzig agglomeration', Geographia Polonica 30: 139-51; J. Czerwinski (1976) 'Problems of protecting the natural environment against the background of economic development in Lower Silesia', Geographia Polonica 34: 273-7; E. Eigendorf and M. Linke (1984) 'The effects of copper ore mining on the cultural landscape of the mansfeld region', Historicka Geografie 23: 143-70; L. Kozacki (1978) 'Changes in the geographical environment as a result of open mining', Geographia Polonica 41: 81-7.
17. F. Probald (1984) 'Urban climate and urban planning in Budapest', in Compton and Pecsi (eds) 191-204.
18. F.W. Carter (1981) 'Conservation problems of historic cities in Eastern Europe', University College London Department of Geography Occasional Paper 39; F.W. Carter (1982) 'Historic cities in Eastern Europe: problems of industrialization, pollution and conservation', Mazingira: International Journal for Environment and Development 6(3): 62-76.
19. T. Sharman (1985) 'Dresden is rebuilt', Geographical Magazine 57: 77-81.
20. A.E.J. Morris (1973) 'Four towns in a Bohemian setting', Geographical Magazine 45: 814-19.

21. A.H. Dawson (1984) 'City profile: Krakow', Cities 1: 449-56.

22. F.W. Carter (1974) 'Concentrated Prague', Geographical Magazine 46: 537-44; F.W. Carter (1984) 'Pollution in Prague: environmental control in a centrally planned socialist economy', Cities 1: 258-73.

23. F.W. Carter (1985) 'Balkan historic cities: pollution versus conservation', in L. Collins (ed.) Anglo-Bulgarian Symposium 1982, London: University of London School of Slavonic and East European Studies, 2: 75-97 (quote p. 76).

CONCLUSION

Eastern Europe has experienced a radical reorganization in the post-war period affecting its political and economic structures. A new territorial structure arose out of the Second World War with particularly radical changes in the northern part of the region. On the other hand there has been absolute stability in boundaries since 1945 and while the strains arising from former political geographies, and from continuing ethnic minority problems in several countries, have not disappeared they are nevertheless contained by the discipline imposed by monopoly communist governments and by the strong powers of co-ordination imposed on the majority of governments by the USSR.

While Russian influence over parts of Eastern Europe has been evident in the past (a large part of present-day Poland was included in the Russian Empire for more than a century before 1918) it has never applied to the whole of the region under review in this book. Now the influence is spatially more extensive and, being Marxist-Leninist in inspiration, involves not only the idealizing of a monopoly party system but equal priority for state ownership of the means of production and central planning of the economy in relation to each country's own resources; involvement in an integrated economy for the bloc, and even more so the global pattern of specialization, has carried only limited priority until recently. The result has been a remarkable change from an agricultural to an industrial emphasis backed up by intensive working of domestic raw materials and fuels (especially low-grade coal), by changes in settlement patterns as young people have left the land and migrated, often on a permanent basis, to the towns, and by the harnessing of domestic, Russian, and increasingly western technology. Differences between countries remain, given the prime emphasis on the nation state, but very high rates of growth have been achieved throughout the region.

Such great changes in their wake naturally give rise to curiosity about the essential ingredients for such revolutionary upheavals. The political and economic elements of East European Socialism are complementary but they are not inseparable and it cannot be assumed that rapid economic development would not have occurred irrespective of political arrangements. It has yet to be convincingly demonstrated that the broad rationale of economic development departs significantly from western practice. Certainly geographers have recognized the same basic problems throughout the continent: while the ecologists express concern over the threat to the environment the spatial scientists find that centre-periphery antagonisms are by no means restricted to western 'capitalist' countries. Regional variations have been held in check by improvements in infrastructure (transport and power) but imbalances in raw materials and skilled labour must be recognized by central planners who seek a high rate of growth nationally. Changes in the system of administrative regions have helped planners to direct capital to compelling resources in areas previously obscured by the demands of large cities under whose authority the backward areas were subordinated. But the large scale of modern factory working, the efficiency involved in developing an integrated industrial estate, and the feasibility of assembling a large labour force by permanent rural-urban migration and daily/ weekly commuting all continue to conspire against the expansionist ambitions of many small towns and key villages. Geographers in Eastern Europe have taken considerable interest in the regional problem and try to discharge their professional responsibility by competing with economists for access to the leading decision-makers in party and government.

Eastern Europe shows dramatic centralizing tendencies, perhaps arising inevitably from central planning related to very large agricultural and industrial organizations and an infrastructure which is better organized nationally (integrating the major cities) than it is regionally and locally. Arguably therefore, to a greater extent than in the west there is a reluctance to establish industries in small towns and to farm marginal areas given the distance from the centre of administration and the additional problems arising from poor transport and power supply. At the same time the system of central planning, typically involving major choices over sector and regional allocations relevant to resources and market requirements (nationally and internationally) for periods of five years or more, creates an element of rigidity in the economy which makes for complicated day-to-day operation as well as inability to adjust. The situation is all the more complex because the vested political interests in the decision-making process make it very difficult to alter course. Eastern Europe therefore displays abundant evidence of economically irrational decision-making which could not be found in the west to the same extent. The enormous costs of centralization and inefficient decision-making is borne by the population in terms of long working hours, low wages, and shortages of everyday requirements. Compared with the west the service sector is poorly developed, given similar levels of agricultural and industrial output. Inevitably 'the traditional system has been challenged for its failure to live up to ideals of socialism and for the dehumanizing effect of the industrialisation rush'. (1) But the situation is exacerbated by an ambivalent attitude by governments towards their people: prosperity can result in greater legitimacy for governments, yet propaganda demands for self-sacrifice in the construction of socialism remain important in recruiting an ideologically motivated community of party members who are essential, along with the military and security organizations, for maintaining the system.

The problem is complicated by the early post-war experience with excessive and premature expansion of manufacturing related to Soviet needs and Soviet fuels and raw materials. This strategy produced rapid structural change in the 1950s and 1960s but it is difficult now to exploit the potential for co-operation with firms in the west and specialization within Eastern Europe. During the 1970s and 1980s there has been less structural change in Eastern Europe than in the west. But changes in the future will arguably bring Eastern and Western Europe closer together. It is no longer possible to maintain a high rate of growth by heavy investment working up to as much as 40 per cent of GNP. To maintain growth the socialist economies must become more efficient and this in turn requires a better use of resources, better management and planning, and a work-force motivated by a clear link between efforts and rewards. All this suggests higher living standards and further economic reform. But when one contemplates the recasting of the central planning system or the break-up of collective farms there are enormous political problems involving vested interests and also Soviet sensibilities. For example, greater efficiency and productivity have serious implications for job security, a socialist principle which governments have been reluctant to ignore (perhaps in part because it helps to maintain the integrity of central planning despite falling growth rates). Major structural changes are needed in the economies of Eastern Europe but these changes cannot really be accomplished without some radical change in the system itself. Economic reform became the slogan of the decade in the 1960s yet even when the changes were superficial the mere allusions to reform constituted a significant gesture by the traditionalists to the economic realities. (2) For despite the propaganda emphasis on revolution the East European system is highly conservative and reforms have been tentative in the extreme. 'We have yet to see a welfare-oriented, consistent and coherent reform, accompanied by a revision of the growth strategy and by a system of checks and balances to curb the omnipotence of the system's directors.' (3)

THE WAY AHEAD

As Eastern Europe has lost its competitiveness in international trade the search for effective solutions is heavily constrained by the Soviet Union's cautious attitude to reform. All the socialist countries of Eastern Europe (except Yugoslavia) now perceive a sharp conflict between the desire for closer links with western industrialized nations and the need for Soviet economic assistance which is of more immediate value to them. Eastern Europe is almost certainly an economic liability to the Soviet Union at the present time but since the Russian interest in the region is essentially political there is a willingness to pay an economic price even though it sustains higher living standards than are available to most citizens of the USSR. (4) However if the Soviet Union wishes at the same time to reduce tension in the world and to improve access to western technology then she must observe certain conventions of civilized behaviour in dealing with her allies.

309

Here it is evident that 'normalization' in Czechoslovakia, for all its bitter disappointments in the eyes of many Czechoslovaks, did not involve the violent arbitrary actions characteristic of Stalinist repression. Thus a balance was struck and the opposition to radical reform in Czechoslovakia was not so insensitive as to prevent the formulation of an Ostpolitik by the Brandt government in the FRG in 1969, a move which the Russians encouraged her allies to reciprocate at a meeting of Warsaw Pact leaders in Moscow at the end of that year. So it may be that the challenge over human rights can be absorbed. There is certainly a democratic tradition, especially in the north, and given the effective parliamentary activity in Eastern Europe (and the scope for individual initiative in business, literature, and religion) before the communist revolutions it is inappropriate to claim that the early post-war experience precludes the development of democratic institutions indefinitely. But for many people the limited progress already made (resulting in greater freedom of speech in Eastern Europe than the Soviet Union) and limited change promised for the future (involving for example a revitalization of parliamentary life in Hungary and Poland) may be acceptable.

But while greater democracy could lead to more pronounced 'differentiation' between the policies of the Soviet Union and those of the East European states, it could also result in more strained relations within the region. Nationalism is becoming stronger. Hungary, for example, has been swept by a wave of 'Habsburg nostalgia' with people openly expressing positive romantic feelings about the 'time of peace' between 1871 and 1914 when war was averted without potentially dangerous nuclear strategies. Yet nationalism can easily sharpen territorial disputes with four most sensitive issues (Kosovo, Macedonia, the North Sea coast, and Transylvania) involving all eight East European states except for Czechoslovakia. In addition there is the problem of the Soviet Union's western frontier: a sensitive issue in Poland and Romania especially. More open discussion could also be dangerous in the context of domestic affairs since the East European states are still not truly homogenous. The migration process, operating through history, has given rise to ethnic and religious divisions while the development of modern national movements in a largely agrarian context creates the possibility that free expression could take counter-modernizing forms with support for the traditional life-styles which presently come most strongly from Moslem peoples like the Albanians of Kosovo. It may therefore be in the interest of all communist governments in the bloc to retain a large measure of control.

Despite the progress made in superpower negotiations in the late 1980s it would be premature to suppose that greater western penetration of Eastern Europe (through an enhanced economic contribution by the FRG or a more powerful cultural presence by the USA) will alter the reality of the rival military systems. It is distinctly fanciful for right-wing analysts to postulate popular support for nationalist anti-communist movements in Comecon countries and in the constituent republics of Yugoslavia to the point where a new state system, guaranteed by the USA, would stand as a barrier to Soviet expansionism. Equally academic are the critical thinkers in Eastern Europe who do not seek a restoration of capitalism but rather a genuinely democratic framework for socialism which could embrace both Eastern and Western Europe. Very important here is the peace movement which looks forward to a dismantling of military alliances, the dissipation of great power influences, and the creation of an environment in which the reunification of Germany as a non-aligned state could then take place. Needless to say such sentiments are regarded with mixed feelings by the authorities in Eastern Europe. Peace movements are to be supported in the west but they must be regarded as subversive at home. Yet opposition must be restrained otherwise the authorities place themselves in the untenable position of failing to discriminate between those 'enemies of socialism' who seek return to western-style capitalism and those activists who advocate a more democratic socialism as provided for by the constitutions of the East European states at the present time. A policy of restraint is required in the hope that political activities can be brought under the control of party organizations or churches which have established good working relations with the authorities.

Economic problems may be more serious. Since East Europeans greatly value the modest improvements to their living standards which they have secured since the early years of communist rule any moves to reduce food consumption through smaller deliveries and higher prices have been highly unpopular. It is evident that the reform movement in Czechoslovakia arose in no small measure as a result of economic stagnation and it is evident that lack of material progress is a critical factor in creating instability to the point when people become impatiently resentful of the need to compromise national interests to suit Soviet security requirements. The one major event in Eastern Europe since 1968, the rise of the reform movement in Poland, was triggered off by economic failures in 1970 when proposed price rises led to rioting and the downfall of the Gomulka regime. Attempts by the government of E. Gierek to increase prices were also frustrated: opposition in 1976 led to the withdrawal of the proposed increases and in 1980 more determined action led to Gierek's removal from power and the formation of a free trade union 'Solidarity'. In 1981 a state of martial law was imposed, the trade union was suppressed and increases in food prices of up to 400 per cent were imposed. (5)

Disillusionment in Romania is becoming profound on account of the economic difficulties combined with an ideological drive for self-sacrifice which has lost much of its credibility. The country remains hard at work and major public works minimize unemployment but with high investment the economy cannot provide an improvement in living standards necessary to maintain a strong sense of regime legitimacy. (6) In assessing the performance of Romania N. Burakow concedes that there has been rapid economic growth and attention to basic human needs, organized on a basis of self-reliance. However 'without proper limits and controls restrictions on human rights and personal freedom can easily extend beyond any temporary measures which could be justified' and furthermore once basic human needs have been satisfied 'the model has not exhibited the ability or the necessary flexibility to provide further significant improvements in the standard of living'. (7) The problems are most acute in Poland and Romania with absolute declines in consumption in the early 1980s but they are symptomatic of the East European system which, despite its promises, finds great difficulty in meeting even the more modest consumer demands over housing, food, and transport.

Writing in 1988 during a period of reform in the Soviet Union it is evident that feelings in Eastern Europe indicate caution and misgiving rather than euphoria. Even if there are none of the miscalculations which have in the past destroyed radical movements in Eastern Europe over-estimating Moscow's capacity for accommodation (the 'Prague Spring' of 1968 being the clearest example) there are profound uncertainties. Economic reform in the Soviet Union will almost certainly result in greater pressure on Eastern Europe to produce better quality goods. The heavy dependence that was conditioned by Soviet military supremacy at the end of the Second World War (and expressed through sweeping institutional changes and the formation of joint companies) is maintained by the co-ordinating role of Comecon and also by the rising cost of Soviet oil (from 1981) which has increased Eastern Europe's exports to the Soviet Union and reduced the scope for trade with the west and with the Third World at a time when political conditions might appear conducive to development. Hard currency debts have become more acutely embarrassing and the dramatic reappraisal of western trading links (converting a 3.1 billion dollar trade deficit in 1980 into a 5.1 billion surplus two years later) has been a major component of the crises already described. Western sanctions (against Poland in retaliation for the suppression of the Solidarity union) have merely served to heighten dependency on the Soviet Union. Western technology will certainly remain very much in demand but perhaps the lessons of the 1970s point to Eastern Europe's capacity for no more than gradual assimilation in sympathy with a hesitant reform process. (8) For even the pace-setting Hungarians face the acute danger of becoming overwhelmed by debts as the propensity to borrow runs ahead of the readiness to compromise such basic tenets of socialism as full employment. The mood is thus one of acute uncertainty and it remains to be seen how far glasnost in the Soviet Union can raise spirits in Eastern Europe. Maximizing economic growth in association with western technology while having proper regard for the constraints of the socio-economic system which the Russians seem likely to insist on retaining remains the key problem for the years ahead.

NOTES

1. G.R. Feiwel (1982) 'A socialist model for economic development: the Polish and Bulgarian experience', in C.K. Wilber and K.P. Jameson (eds) Socialist Models of Development, Oxford: Pergamon, 929-50 (quote p. 947).

2. G. Grossman (ed.) (1970) Essays in Socialism and Planning in Honour of Carl Landauer, Englewood Cliffs, NJ:

Prentice-Hall.

3.   G.R. Feiwel (1982) 'A socialist model for economic development: the Polish and Bulgarian experience', in C.K. Wilber and K.P. Jameson (eds) Socialist Models of Development, Oxford: Pergamon, 947.

4.   P. Marer (1976) 'Has Eastern Europe become a liability to the Soviet Union?', in C. Gati (ed.) The International Politics of Eastern Europe, New York: Praeger, 59-80.

5.   P. Summerscale (1982) The East European Predicament: Changing Patterns in Poland Czechoslovakia and Romania, Aldershot: Gower. See also Bahro (1978); J.F. Brown (1977) Eastern Europe's Uncertain Future, New York: Praeger; T. Gilberg (1985) 'Eastern Europe at the crossroads: contradictory tendencies of subservience and autonomy', in S. Bethlen and I. Volgyes (eds) Europe and the Superpowers: Political Economic and Military Policies in the 1980s, Boulder, Col: Westview; P. Lewis (ed.) (1984) Eastern Europe: Political Crisis and Legitimation, London: Croom Helm; M.J. Sodaro and S.L. Wolchik (eds) (1983) Foreign and Domestic Policy in Eastern Europe in the 1980s: Trends and Prospects, London: Macmillan; A.B. Ulam (1974) 'The destiny of Eastern Europe', Problems of Communism 23(1): 1-12.

6.   M.E. Fischer (1977) 'Participatory reforms and political development in Romania', in J.F. Triska and P.M. Cocks (eds) Political Developments in Eastern Europe, New York: Praeger, 217-37.

7.   N. Burakow (1982) 'Romania and Greece: socialism versus capitalism', in C.K. Wilber and K.P. Jameson (eds) Socialist Models of Development, Oxford: Pergamon, 907-28 (quote p. 924).

8.   T.W. Simons (1988) 'Approaching relations with Eastern Europe in the late 1980s', Wilson Center East European Program Occasional Paper 12. See also J.P. Hardt and C.H. McMillan (1988) Planned Economies: Confronting the Challenge of the 1980s, Cambridge: Cambridge University Press.

BIBLIOGRAPHY

References on East European political, demographic, and economic themes will be found in the companion volume. The attention of readers is drawn to two German atlases: the work of T. Kraus et al. (1959) Atlas ost iches Mitteleuropa, Bielefeld: Velhagen & Klasing, is excellent for the GDR, Poland, and northern Czechoslovakia while remaining parts of Eastern Europe feature in J. Breu (1970) Atlas der Donaulander, Vienna: Osterreichisches Ost- und Sudosteuropa Institut. National atlases are available for Bulgaria (1973), Czechoslovakia (1966), and Romania (1979).

Ambler, J. et al. (eds) (1985) Soviet and East European Transport Problems, London: Croom Helm.

Ausch, S. (1972) Theory and Practice of CMEA Cooperation, Budapest: Hungarian Academy of Sciences.

Bahro, R. (1978) The Alternative in Eastern Europe, London: New Left Books.

Bencze, I. and Tajti, E.V. (1972) Budapest: An Industrial Geographical Approach, Budapest: Hungarian Academy of Sciences.

Benes, V.L. and Pounds, N.J.G. (1970) Poland, London: Benn.

Bertsch, G.K. and McIntyre, J.R. (eds) (1983) National Security and Technology Transfer: The Strategic Dimension of East-West Trade, Boulder, Col: Westview.

Bethlen, S. and Volgyes, I. (eds) (1985) Europe and the Superpowers: Political Economic and Military Policies in the 1980s, Boulder, Col: Westview.

Bland, W.B. (1988) World Bibliographical Series: Albania, Oxford: Clio.

Biberaj, A. (1988) Albania: a nation in transition, Boulder, Col: Westview.

Bolitho, A. (1971) Foreign Trade Criteria in Socialist Countries, Cambridge: Cambridge University Press.

Bornstein, M. et al. (eds) (1981) East-West Relations and the Future of Eastern Europe, London: Allen & Unwin.

Bourne, L.S. et al. (eds) (1984) Urbanization and Settlement Systems: International Perspectives, Oxford: Oxford University Press.

Brada, J.C. (ed.) (1976) Quantitative and Analytical Studies in East-West Relations, Bloomington, Ind: Indiana University Press.

Bromke, A. (1985) Eastern Europe in the Aftermath of Solidarity, Boulder, Col: East European Monographs.

Brown, A. and Neuberger, E. (eds) (1968) International Trade and Central Planning, Berkeley, Calif: University of California Press.

Brown, A. et al. (eds) (1974) Urban and Social Policies in Planned and Market Economies, New York: Praeger.

Brown, J.F. (1977) Eastern Europe's Uncertain Future, New York: Praeger.

Bryson, P.J. (1984) The Consumer under Socialist Planning: The East German Case, New York: Praeger.

Burghardt, A.F. (ed.) (1975) Development Regions in the Soviet Union Eastern Europe and Canada, New York: Praeger.

Childs, D. (1971) Germany since 1918, London: Batsford.

Coker, C. (1985) NATO, the Warsaw Pact and Africa, London: Macmillan.

Compton, P.A. and Pecsi, M. (eds) (1984) Environmental Management: British and Hungarian Case Studies, Budapest: Hungarian Academy of Sciences.

Connor, W.D. (1979) Socialism Politics and Equity, New York: Columbia University Press.

Davies, N. (1986) A short history of Poland, Oxford: Oxford University Press.

Dawisha, K. and Hanson, P. (eds) (1981) Soviet-East European Dilemmas, London: Holmes & Meier.

Dawson, A.H. (ed.) (1986) Planning in Eastern Europe, London: Croom Helm.

Deletant, A. and Deletant, D. (1985) World Bibliographical Series: Romania, Oxford: Clio.

Demek, J. et al. (1971) Geography of Czechoslovakia, Prague: Academia.

Demko, G. (ed.) (1984) Regional Development Problems and

Policies in Eastern and Western Europe, London: Croom Helm.

Dewar, M. (1951) Soviet Trade with Eastern Europe, London: RIIA.

Dobb, M. (1970) Socialist Planning: Some Problems, London: Lawrence & Wishart.

Drachkovitch, M. (1982) East Central Europe: Yesterday Today Tomorrow, Stanford, Calif: Hoover Institution Press.

Enyedi, G. (ed.) (1976a) Rural Transformation in Hungary, Budapest: Hungarian Academy of Sciences.

----- (1976b) Hungary: An Economic Geography, Boulder, Col: Westview.

Fallenbuchl, Z.M. (ed.) (1975) Economic Development in the Soviet Union and Eastern Europe, New York: Praeger.

Fedorowicz, J.K. (1986) East-West Trade in the 1980s: Prospects and Policies, Boulder, Col: Westview.

Fischer-Galati, S. (ed.) (1963) Eastern Europe in the Sixties, New York: Praeger.

----- (1969) The Socialist Republic of Romania, Baltimore, Md: Johns Hopkins University Press.

----- (1970) Twentieth-Century Romania, New York: Columbia University Press.

----- (ed.) (1981) Eastern Europe in the 1980s, Boulder, Col: Westview.

Fisher, J.C. (1966) Yugoslavia: A Multinational State, San Francisco: Chandler House.

----- (ed.) (1966) City and Regional Planning in Eastern Europe, Ithaca, NY: Cornell University Press.

French, R.A. and Hamilton, F.E.I. (eds) (1979) The Socialist City: Spatial Structure and Urban Policy, Chichester: Wiley.

Fullenbach, J. (1981) European Environmental Policy: East and West, London: Butterworth.

Gati, C. (ed.) (1974) The Politics of Modernization in Eastern Europe, New York: Praeger.

----- (ed.) (1976) The International Politics of Eastern Europe, New York: Praeger.

Georgescu, V. (ed.) (1985) Romania: Forty Years 1944-1984, New York: Praeger.

Gilberg, T. (1975) Modernization in Romania since World War Two, New York: Praeger.

Halpern, J. (1972) A Serbian Village in Historical Perspective, New York: Holt, Rinehart, & Winston.

Hamilton, F.E.I. (1975) Poland's Western and Northern

Territories, Oxford: Oxford University Press.

Hanhardt, A.M. (1986) German Democratic Republic, London: Frances Pinter.

Hann, C.M. (1980) Tazlar: A Village in Hungary, Cambridge: Cambridge University Press.

---- (1985) The Village without Solidarity: Polish Peasants in Years of Crisis, New Haven, Conn: Yale University Press.

Hansen, N.M. (ed.) (1978) Human Settlement Systems: International Perspectives on Structure Change and Public Policy, Cambridge, Mass: Billinger.

Haydon, E.W. (1976) Technology Transfer to Eastern Europe: United States Corporate Experience, New York: Praeger.

Heinrich, H-G. (1986) Hungary: Politics Economics and Society, London: Frances Pinter.

Hill, M.R. (1983) East-West Trade: Industrial Co-operation and Technology Transfer: The British Experience, Aldershot: Gower.

Hoffman, G.W. (ed.) (1971) Eastern Europe: Essays in Geographical Problems, London: Methuen.

---- (1972) Regional Development Structure in Southeastern Europe, New York: Praeger.

Hoffman, G.W. and Dienes, L. (1985) The European Energy Challenge: East and West, Durham, NC: Duke University Press.

Holzman, F.D. (1987) The Economics of Soviet Bloc Trade and Finance, Boulder, Col: Westview.

Ignotus, P. (1972) Hungary, London: Benn.

Jackson, M.R. and Woodson, J.D. (eds) (1984) New Horizons in East-West Economic and Business Relations, Boulder, Col: Westview.

John, L.G. (ed.) (1975) EEC Policy towards Eastern Europe, Farnborough: Saxon House.

Joseph, P. (ed.) (1987) The Economies of Eastern Europe and their Foreign Trade Relations, Brussels: NATO Economic Directorate.

Kabdebo, T. (1980) World Bibliographical Series: Hungary, Oxford: Clio.

Kalla-Bishop, P.M. (1973) Hungarian Railways, Newton Abbot: David & Charles.

Kanet, R.E. (1988) The Soviet Union Eastern Europe and the Third World, Cambridge: Cambridge University Press.

Kansky, K. (1976) Urbanization under Socialism: The Case of Czechoslovakia, New York: Praeger.

Kaser, M. (1967) Comecon: Integration Problems for the
    Planned Economies, Oxford: Oxford University Press.
---- (ed.) (1968) Economic Development for Eastern Europe,
    London: Macmillan.
Kende, P. and Strmiska, Z. (1987) Equality and Inequality in
    Eastern Europe, Leamington Spa: Berg.
Kiss, T. (1971) International Division of Labour in Open
    Economies with Special Regard to the CMEA, Budapest:
    Hungarian Academy of Sciences.
Kohler, H. (1965) Economic Integration in the Soviet Bloc
    with an East German Case Study, New York: Praeger.
Kolankiewicz, G. and Lewis, P.G. (1988) Poland: Economics
    and Society, London: Frances Pinter.
Kosinski, L.A. (ed.) (1977) Demographic Developments in
    Eastern Europe, New York: Praeger.
Kostanick, H.L. (ed.) (1977) Population and Migration Trends
    in Eastern Europe, Boulder, Col: Westview.
Krisch, H. (1985) The GDR: The Search for Identity,
    Boulder, Col: Westview.
Kuhlman, J.A. (ed.) (1978) The Foreign Policies of Eastern
    Europe, London: Sijthoff.
Kuklinski, A. (ed.) (1977) Regional Studies in Poland,
    Warsaw: Polish Academy of Sciences.
Lampe, J.R. (1986) The Bulgarian Economy in the Twentieth
    Century, London: Croom Helm.
Levcik, F. and Stankovsky, J. (1979) Industrial Co-operation
    between East and West, London: Macmillan.
Lewanski, R.C. (1984) World Bibliographical Series: Poland,
    Oxford: Clio.
McFarlane, B. (1988) Yugoslavia: Politics, Economics and
    Society, London: Frances Pinter.
McIntyre, R.J. (1986) Bulgaria, London: Frances Pinter.
Marer, P. (1972) Soviet and East European Trade 1946-1969:
    Statistical Compendium and Guide, Bloomington, Ind:
    Indiana University Press.
Marmallaku, R. (1975) Albania and the Albanians, London:
    Hurst.
Masnata, A. (1974) East-West Economic Co-operation:
    Problems and Solutions, Aldershot: Gower.
Matley, I.M. (1970) Romania: A Profile, London: Pall Mall.
Mellor, R.E.H. (1971) Comecon: Challenge to the West, New
    York: Van Nostrand.
---- (1975) Eastern Europe: A Geography of the Comecon
    Countries, London: Macmillan.
---- (1978) The Two Germanies: A Modern Geography,

London: Harper & Row.

Mieczowski, B. (ed.) (1980) Eastern European Transportation: Systems and Modes, The Hague: Nijhoff.

Mihailovic, K. (1972) Regional Development: Experiences and Prospects in Eastern Europe, The Hague: Mouton.

Musil, J. (1981) Urbanization in Socialist Countries, London: Croom Helm.

Musoroiu, M. (ed.) (1974) Urban Growth Processes in Romania, Bucharest: Meridiane.

Narkiewicz, O. (1986) Eastern Europe 1968-1984, London: Croom Helm.

Nelson, D.N. (ed.) (1981) Romania in the 1980s, Boulder, Col: Westview.

Osborne, R.H. (1967) East Central Europe, London: Chatto & Windus.

Paliwoda, S.J. (1981) Joint East-West Marketing and Production Ventures, Aldershot: Gower.

Pano, N.C. (1968) The People's Republic of Albania, Baltimore, Md: Johns Hopkins University Press.

---- (1986) Albania, London: Frances Pinter.

Paul, D.W. (1981) Czechoslovakia: Profile of a Socialist Republic at the Crossroads of Europe, Boulder, Col: Westview.

Pavlowitch, S.K. (1971) Yugoslavia, London: Benn.

---- (1988) The Improbable Survivor: Yugoslavia and its Problems 1915-1988, London: Hurst.

Patyk, J.F. (1966) The CMEA: Its Political Importance, University of Colorada: Ph.D. thesis.

Pecsi, M. (ed.) (1964) Applied Geography in Hungary, Budapest: Hungarian Academy of Sciences.

Pecsi, M. and Probald, F. (eds) (1974) Man and Environment, Budapest: Hungarian Academy of Sciences.

Pecsi, M. and Sarfalvi, B. (1964) The Geography of Hungary, London: Collets.

Pounds, N.J.G. (1964) Poland between East and West, Princeton, NJ: Princeton University Press.

---- (1969) Eastern Europe, London: Longman.

Pryor, F.L. (1963) The Communist Foreign Trade System, London: Allen & Unwin.

Rachwald, A.R. (1983) Poland between the Superpowers: Security versus Economic Recovery, Boulder, Col: Westview.

Radu, M. (ed.) (1981) Eastern Europe and the Third World: East versus South, New York: Praeger.

Ramet, P. (ed.) (1988) Yugoslavia in the 1980s, Boulder, Col: Westview.

Ronnas, P. (1984) Urbanization in Romania: A Geography of Social and Economic Change since Independence, Stockholm: Stockholm School of Economics.

Rubinstein, A.Z. (1970) Yugoslavia and the Non-Aligned World, Princeton, NJ: Princeton University Press.

Rugg, D.S. (1985) The World's Landscapes: Eastern Europe, London: Longman.

Rupnik, J. (1986) Czechoslovakia, London: Frances Pinter.

Sampson, S. (1984) National Integration through Socialist Planning: An Anthropological Study of a Romanian New Town, Boulder, Col: East European Monographs.

Sarfalvi, B. (ed.) (1969) Problems in Hungarian Applied Geography, Budapest: Hungarian Academy of Sciences.

---- (ed.) (1971) The Changing Face of the Great Hungarian Plain, Budapest: Hungarian Academy of Sciences.

---- (ed.) (1975) Urbanization in Europe, Budapest: Hungarian Academy of Sciences.

Saunders, C.T. (ed.) (1980) East and West in the Energy Squeeze: Prospects for Co-operation, London: Macmillan.

---- (ed.) (1981) East-West-South: Economic Interaction between Three Worlds, London: Macmillan.

---- (ed.) (1983) Regional Integration in East and West, London: Macmillan.

---- (ed.) (1985) East-West Trade and Finance in the World Economy, London: Macmillan.

Schaefer, H.W. (1972) Comecon and the Politics of Integration, New York: Praeger.

Schopflin, G. (ed.) (1970) The Soviet Union and Eastern Europe: A Handbook, London: Blond.

Shafir, M. (1985) Romania: Politics Economics and Society, London: Frances Pinter.

Sharman, T. (1988) Poland, London: Columbus Books.

Shlaim, A. and Yannopoulos, G.N. (eds) (1978) The EEC and Eastern Europe, Cambridge: Cambridge University Press.

Short, D. (1986) World Bibliographical Series: Czechoslovakia, Oxford: Clio.

Sillince, J. (ed.) (1989) Housing in Eastern Europe, London: Routledge.

Simic, A. (1973) The Peasant Urbanites: A Study of Rural-Urban Mobility in Serbia, New York: Seminar Press.

Singleton, F.B. (1970) Yugoslavia: The Country and its

People, London: Queen Anne Press.

---- (ed.) (1987) Environmental Problems in the Soviet Union and Eastern Europe, London: Lynne Rienner.

Singleton, F.B. and Carter, B. (1932) The Economy of Yugoslavia, London: Croom Helm.

Sininian. S. et al. (eds) (1972) Eastern Europe in the 1970s, New York: Praeger.

Stern, J.P. (1982) East European Energy and East-West Trade in Energy, London: Policy Studies Institute and RIIA.

Stokke, B.R. (1968) Soviet and East European Trade and Aid in Africa, New York: Praeger.

Tamke, J. (1983) The People's Republics of Eastern Europe, London: Croom Helm.

Taras, R. (1986) Poland: Socialist State Rebellious Nation, Boulder, Col: Westview

Tismer, J.F. et al. (eds) (1987) Transport and Economic Development: Soviet Union and Eastern Europe, Berlin: Osteuropa-Institut an der Frien Universitat Berlin.

Triska, J.F. and Cocks, P.M. (eds) (1977) Political Development in Eastern Europe, New York: Praeger.

Turnock, D. (1974) An Economic Geography of Romania, London: Bell.

---- (1978) Studies in Industrial Geography: Eastern Europe, Folkestone: Dawson.

---- (1980) The Human Geography of the Romanian Carpathians, Nottingham: Geographical Field Group.

---- (1986) The Romanian Economy in the Twentieth Century, London: Croom Helm.

Vajda, I. and Simai, M. (eds) (1971) Foreign Trade on a Planned Economy, Cambridge: Cambridge University Press.

van Brabant, J.M. (1980) Socialist Economic Integration: Aspects of Contemporary Economic Problems in Eastern Europe, Cambridge: Cambridge University Press.

Verdery, K. (1983) Transylvanian Villagers, Berkeley, Calif: University of California Press.

Vine, R.D. (ed.) (1987) Soviet-East European Reforms as a Problem for Western Policy, London: Croom Helm.

Volgyes, I. (ed.) (1974) Environmental Deterioration in the Soviet Union and Eastern Europe, New York: Praeger.

---- et al. (eds) (1980) The Process of Rural Transformation, New York: Pergamon.

---- (1981) Hungary: A Profile, Boulder, Col: Westview.

---- (1982) Hungary: A Nation of Contradictions, Boulder,
    Col: Westview.
Wallace, W.W. (1977) Czechoslovakia, London: Benn.
Wanklyn, H.G. (1954) Czechoslovakia: A Geographical and
    Historical Study, London: Philip.
Wasowski, S. (ed.) (1970) East-West Trade and the
    Technology Gap, New York: Praeger.
Welsh, W.A. (1988) Bulgaria, Boulder, Col: Westview.
Wilczynski, J. (1969) The Economics and Politics of East-
    West Trade, London: Macmillan.
---- (1974) Technology in Comecon: Acceleration of
    Technological Progress through Economic Planning,
    London: Macmillan.
---- (1977) The Economics of Socialism, London: Allen &
    Unwin.
Wiles, P.J.D. (1969) Communist International Economics,
    Oxford: Blackwell.
Woodward, S. (1987) Yugoslavia, London: Frances Pinter.
Zloch-Christy, I. (1988) Debt Problems of Eastern Europe,
    Cambridge: Cambridge University Press.
Zonoviev, A. (1984) The Reality of Communism, London:
    Gollancz.

INDEX